BIOTECHNOLOGY IN AGRICULTURE, INDUSTRY AND MEDICINE

ADVANCES IN BIOTECHNOLOGY

A PRACTICAL APPROACH

BIOTECHNOLOGY IN AGRICULTURE, INDUSTRY AND MEDICINE

Additional books in this series can be found on Nova's website under the Series tab.

Additional e-books in this series can be found on Nova's website under the e-book tab.

BIOTECHNOLOGY IN AGRICULTURE, INDUSTRY AND MEDICINE

ADVANCES IN BIOTECHNOLOGY

A PRACTICAL APPROACH

HARISH KUMAR DHINGRA
P. N. JHA
AND
P. BAJPAI
EDITORS

New York

For permission to use material from this book please contact us:
Telephone 631-231-7269; Fax 631-231-8175
Web Site: http://www.novapublishers.com

Library of Congress Cataloging-in-Publication Data

ISBN: 978-1-62417-842-9

Published by Nova Science Publishers, Inc. † New York

CONTENTS

PREFACE

Biotechnology- the use of living organisms to enhance or their products to enhance our lives and our environment- is a broad and complex discipline that encompasses many specialized areas. The promise that biotechnology holds for developing countries is well recognized. An organized loom aimed at the applications of biotechnology in a variety of sectors would prove beneficial for national development. Biotechnology is an important tool that can be applied to diverse economic sectors of special interest to developing countries, such as the production of food crops, livestock management, human health care, chemical industries and environmental management.

The emergence of biotechnology over the last two decades as a major discipline has opened up new landscapes for increased productivity on all fronts-agriculture, medicine, and industry, particularly in relevance to developing countries confronted with an ever increasing population, food shortage, and scarcity of economic resources.

The treatment of subjects is necessarily selective but the book seeks to balance the traditional with the new biotechnologies and science with their applications and potential.

The chapters of the book have been contributed by distinguished experts in the field. The book covers applications of biotechnology in selected areas such as health care, agriculture, microbial system, "in silico" analysis for drug designing and drug discovery and the environment.

ABOUT THE EDITORS

Dr. Harish kumar Dhingra is working as Assistant Professor in the Depatrment of Science, Faculty of Arts, Science and Commerce, Mody Institute of Technology & Science (Deemed University), Laxmangarh, Distt. Sikar, Rajasthan. He has contributed many research articles in the field of Biotechnology. He has published many chapters in various books. He is having five year experience of teaching to various UG, PG, & Ph.D. students. His interenst lies in inspiring students in the elaborated horizon & scope of Biotehcnology.

Dr. Prabhat Nath Jha is working as Assistant Professor in the Depatrment of Biological Sciences, Birla Institute of Technology & Science (BITS), Pilani, Rajasthan. He is having five year experience of teaching to various UG, PG, & Ph.D. students. He has published a number of national and international research papers.

Ms. Pratima Bajpai is working as Assistant Professor in the Depatrment of Science, Faculty of Arts, Science and Commerce, Mody Institute of Technology & Science (Deemed University), Laxmangarh, Distt. Sikar, Rajasthan. She has published many chapters in various books.

ACKNOWLEDGEMENT

First of all, I would like to express my sincere gratitude to the almighty that has made me capable for preparing this manuscript.

The completion of this book has come through the overwhelming help that came from many people whom mentioning each one may not be exhaustive. I wish to express my sincere gratitude to all the people who offered their kind help and guidance all the time till completion of this manuscript.

I deeply appreciate the efforts of co-editors of this manuscript, Dr. Prabhat Nath Jha, Assistant Professor, Dept. of Biosciences, Birla Institute of Technology and Science, Pilani and Ms. Pratima Bajpai, Assistant Professor in Science, Mody Institute of Technology & Science, Laxmangarh, Distt. Sikar, Rajasthan whose help, guidance and experience, has assisted me in production of this book.

I would also like to thank all the authors of different chapters included in this book for providing me the data of this manuscript.

My sincere acknowledgements are due to Prof. Dr. Suneel Jakhoria, Dean FASC, MITS University, for her kindness and inspiration through out the course.

My special gratitude also goes to R. K. Gaur (HOD, Department of Science, FASC, MITS University) and Dr. Narendra Kumar, Faculty, FASC, MITS for their immense support and help.

My heart felt gratitude to all my friends who helped me in completing this project. Last but not least, I thank all my family members, especially my father, for all the support during completion of this manuscript.

Dr. Harish Kumar Dhingra

In: Advances in Biotechnology
Editors: H. K. Dhingra, P. N. Jha et al.
ISBN: 978-1-62417-842-9

Chapter 1

GENOME ANALYSIS USING A COMPUTATIONAL APPROACH

Satendra Singh[1,*], Prashant Ankur Jain[1], Budhayash Gautam[1], Atul Kumar Singh[2], Chandrabhan Seniya[2] and G. J. Khan[3]

[1]Department of Computational Biology and Bioinformatics, SHIATS, Allahabad, India

[2]Department of Biotechnology, Madhav Institute of Technology and Science, Gwalior, M.P., India

[3]Department of Biotechnology, L.N. Mithila University, Darbhanga, Bihar, India

INTRODUCTION

Recent advances in DNA sequencing technology have enabled elucidation of whole genome information from a plethora of organisms. In parallel with this technology, various bioinformatics tools have driven the comparative analysis of the genome sequences between species and within isolates. While drawing meaningful conclusions from a large amount of raw material, is the main task that comes under genome analysis. The genomic data is too large and complex that we cannot extract meaningful information without using

* Corresponding Author: Satendra Singh, Department of Computational Biology and Bioinformatics, SHIATS, Allahabad-211007, India. E-mail:satendralike@gmail.com.

computation methods and tools. In parallel with the genome projects, a number of bioinformatics tools have been developed to facilitate *in silico* analysis of the gene sequence information. A major application of Bioinformatics is analysis of the full genomes of organisms that have been sequenced starting in the late 1990s, including microbial genomes, the budding yeast *Saccharomyces cerevisiae*, the nematode worm *Caenorhabditiselegans,* the plant *Arabidopsis thaliana*, the fruit fly *Drosophila*, and the human genome. Many additional genome sequencing projects are either being planned or are already under way. Traditional genetics and molecular biology have been directed toward understanding the role of a particular gene or protein in an important biological process. A gene can be sequenced and computational methods can be used to predict its function or to manipulate its activity or expression. In contrast the availability of genome sequences provides the sequences of all the genes of an organism so that important genes influencing metabolism, cellular differentiation and development, and disease processes in animals and plants, can be identified and the relevant genes manipulated [1].

The challenge is to identify those genes that are predicted to have a particular biological function and then to design experiments to test that prediction. This analysis depends on gene prediction using gene models for each organism followed by sequence comparisons between the predicted proteins with other proteins whose function is known from the biological studies. To facilitate such comparisons, the genomes of a number of model organisms about which a great deal of biological information is available have been sequenced. Many years of genetic and biochemical research of these model organisms such as *Escherichia coli, S. cerevisiae,C. elegans,A. thaliana,* and *D. melanogaster* have led to the accumulation of a large amount of information on gene organization and function. The mouse *Musmusculus* is a genetic model for humans because the two species are so closely related through evolution.

A newly identified gene in another organism can be compared to the existing database of information to find whether it has a similar function. Genes involved in human disease, for example, are sometimes found to be similar to a fruit fly gene at the protein sequence level. The genetic effects of mutations in the fruit fly's gene will then provide a biochemical, cellular, or developmental model for the human disease. Interestingly, it has not been possible to identify the function of all the genes in model organisms. As a result, a similar gene or family of genes may be found in several organisms, including a model organism, but the function is not known because the gene functions have not yet been analyzed. Hence, continued biological analysis of model organisms in those areas that are not tractable by the tools of Bioinformatics has many important applications. Tracing the Phylogenitic history of such uncharacterized genes, characterized genes and gene domains and gene linkages in diverse organisms is one of the most interesting and challenging aspects of genome analysis [2]. In addition, even though a gene that specifies an important biological function has not been identified, the gene can be traced in individuals using sequence variations that occur among individuals in a population, called sequence polymorphisms. In humans, for example, single nucleotide polymorphisms (SNPs) can be found throughout the genome, including some that are positioned adjacent to an important disease gene. The computational tools can be used to extract information from SNPs that can be used in pharmacogenomics for personalized or individual therapy.

Hemophilusinfluenzae was the first bacterial genome to be sequenced by Genomics Research (TIGR, http://www.tigr.org) using the shotgun method. Since then the availability of genome sequences greatly facilitates the discovery and utilization of these sequence polymorphisms. It is recognized that some types of genetic variation, including specific human diseases, are best understood at the genome-wide level. The duplication of genes, gene segments, and gene clusters provides opportunities for recombination events that can cause changes in gene copy number or loss of gene function.

Computational Approaches for Genome Analysis

Prokaryotic genome has some different features in comparison to that of eukaryotic cells. In the following sections some of the genome features are discussed along with their significance in genome analysis. Once the sequence is available, open reading frames are identified, and these are compared to the existing proteins by a database similarity search [3]. In examining the results of such analysis, it is important to look for the method which uses the statistical significance of the result, and overall degree of confidence in the alignments. The analysis should be repeated if necessary (Figure 1).

The prokaryotic genome analysis is comparatively less complex than eukaryotic genome analysis as the eukaryotic genome size is generally larger and has complicated genome structure. For example the eukaryotic genome is composed of tandem repeats of sequences of varying length, the genes are composed of both intron and exons etc.

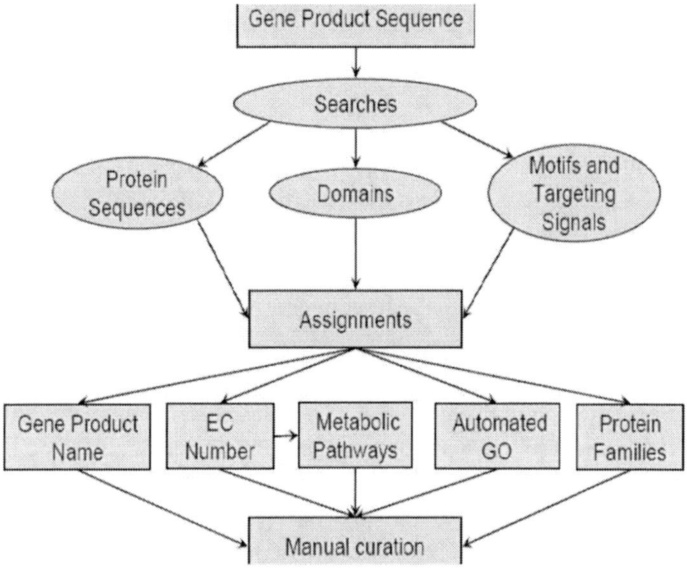

Figure 1. Various approaches for genome analysis.

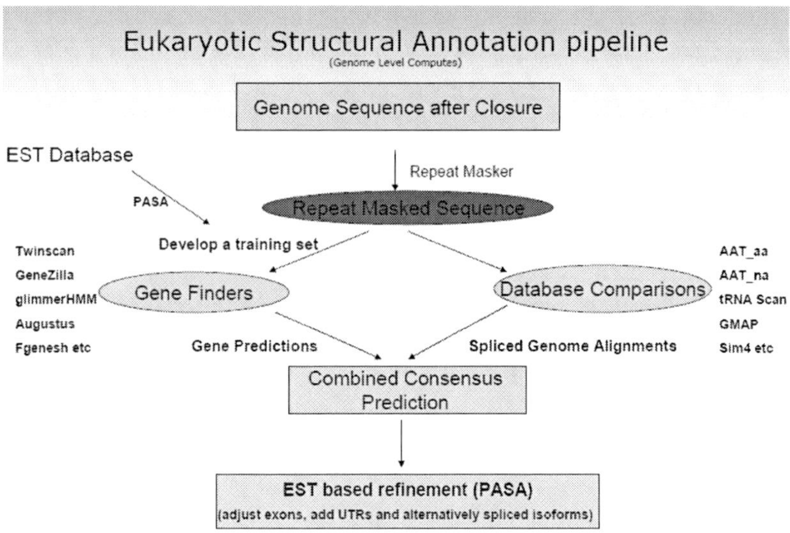

Figure 2. Pipeline for eukaryotic genome annotation.

On other hand prokaryotic genome contains genes uninterrupted by intron and possesses a few, repeat. Such feature of eukaryotic and prokaryotic genome requires different algorithms to extract the biological information present in the genome (Figure 2). The algorithms designed to handle genomic data utilizes genomic feature for gene prediction in the genome. For example genomic features include the consensus sequences at exon–intron and intron–exon splice junctions, base composition, codon usage, and preference for neighboring codons etc.

REPEATS

The repetitive segment of DNA sequence in the genome forms the sequence repeats. Sequence repeats are the repeats of the same DNA sequence. The sequence repeats are found in genome of different organisms in different size, and at different locations. These elements can comprise a large proportion of the eukaryotic genome as repetitive sequences. The sequence repeats are important from the evolutionary point of view for a genome.

TRANSPOSABLE ELEMENTS

The transposable elements (TEs) are short DNA sequences that can move from one chromosomal location during the evolution of the genome. Transposable elements sometimes covers large portion of genome in both plants and animals. They too play significant role in genome evolution. TEs can comprise one-half or more of the genome sequence. The genes in eukaryote are interrupted by introns, having varying length and number. The introns in the genes of some genome may sometimes comprise only a small fraction of the genes and in other genomes the introns comprise more than 95% of the gene. Intron changes their position in the genes during evolution and thus effects the gene expression [4]. The computation methods use the introns for the gene prediction in the genome.

PSEUDOGENES

The pseudogenes are DNA sequences that were derived from a functional copy of a gene during the evolutionary time period of genome. The pseudogenes generally accumulate the mutations. As they are derived from functional copy of genes having mutations, the function of such pseudogenes gets destroyed. They can be predicted in a genome by using all-against-all search method. Presence of pseudogenes in a organism indicates the possible role of gene duplication in genome expansion and reshaping during evolution.

PREDICTING GENE IN NEWLY SEQUENCED GENOME

The newly sequenced genome sequence is scanned for protein encoding genes. Each predicted gene is translated into a protein sequence. The complete set of protein sequences encoded by the genome is known as the proteome of that organism. Every protein in the proteome is then used as a query sequence in a database similarity search. The validation of the protein prediction can be carried out by screening the predicted protein sequences against an expressed sequence tag (EST) library which can confirm the prediction and expression of the gene [5]. Each validated protein sequence can further be used to identify the paralogous (by searching each protein against rest of the proteome) and orthologous protein (by searching each protein against the proteome of different organisms) (Figure 2).

Genome sequences are generated from the assembled contigs from genomic libraries. The genomic library is constructed by compilation of cloned fragments of DNA sequence. Each chromosome in the genome is fragmented into sort segments, which were separated, purified and cloned for the generation of genomic library. The sequencing of cloned DNA segments can be carried out by methods like chain termination method or any approach of next generation sequencing strategies. The full chromosomal sequences are then assembled from the overlaps in a highly redundant set of fragments by an automatic computational method or from the fragment order on a physical map [6].

Knowing the base pair sequence is only the first step and basic step in the genome analysis. The real task in genome analysis stats once we know the genome sequence. The different genome (eukaryotes and prokaryotes) has different approaches for their analysis, as they have different composition and organization of genome structures. It is relatively easy to analyze and generate information for prokaryotic genome in comparison to the eukaryotic genomes. The Eukaryotic genomes structure is composed of different elements like repeated elements, including tandem repeats and transposable elements which make genome analysis more complex. The sequenced DNA sequence is used to predict gene with the help of computational programs which can further be used to predict the corresponding amino acid sequence of proteins encoded by the predicted genes. The annotation of such predicted proteins can be performed by using protein sequence as query for the databases similarity searches. The alignment score, query coverage and e-value can be used to validate the quality of the alignment in the database search (Figure 1).

Sometimes database searching methods for query sequence fails to answer the expected question about the query sequence, under such condition instead of predicting biological properties based on sequence comparison we can do so by predicting other feature of sequence like, motifs, conserve domain, regular expression etc. These features can give indication about biological properties of the query sequence.

In case we are interested in analyzing overall gene expression of a genome the most reliable and efficient method of choice is microarray data analysis. The microarray data analysis involves the simultaneous analysis of thousand of genes in a single experiment. The over expressed genes, under expressed genes, normally expressed genes and co expressed genes can be easily identified using different clustering algorithms [7]. Such findings can be further extended to predict the function of genes, their identification and role of genes in various metabolic pathways.

Once the predicted genes are identified and annotated, they can be used to establish or construct the various metabolic pathways in organism. The metabolic pathways of various organisms are already known with their detailed activities. These can be used to study the metabolic pathway activities of newly sequenced genome. Each predicted gene is identified and can be used to find its location in the genome. The genome map can be constructed where we can find the relative location of genes in the chromosomes. The similar gene order among organisms reflects evolutionary relationship among the genes. An all-against-all search can be performed where each protein is used as query against rest of proteome of the organism. The significant proteins thus identified are known as paralogous protein. The proteome may be compared to itself to identify paralogs, families of proteins that have arisen by gene duplication. One proteome may also be compared to another proteome to discover orthologous genes that have kept the same function, genes that have become fused to make a larger protein (or split into two to make two separate proteins) new arrangements of protein domains, and amplification of protein families to perform a new type of biological function. The identified paralogous proteins can further be used for the cluster analysis. The proteins grouped in same cluster may share various biological features with each other (Table 1).

Table 1. Different tools/websites/servers/centers for genome analysis

Database or organization	Internet address
Sequence databases	
NCBI	http://www.ncbi.nlm.nih.gov/Entrez/ Genome/org.html
DDBJ	http://www.ddbj.nig.ac.jp/htmls_test/ Welcome-e.html
EBI/EMBL	http://www.ebi.ac.uk/ebi_home.html
GSDB	http://www.ncgr.org/gsdb/Index_gsdb.html
SwissProt (Geneva)	http://expasy.hcuge.ch/www/expasy-top.html
Candida	http://alces.med.umn.edu/Candida.html
MIPs	http://www.mips.biochem.mpg.de/
RDP	http://rdp.life.uiuc.edu/
SGD	http://genome-www.stanford.edu/
Metabolic databases	
KEGG	http://www.genome.ad.jp/kegg/
Ecocyc	http://ecocyc.PangeaSystems.com/ ecocyc/ecocyc.html
WIT	http://www.cme.msu.edu/WIT/ Sequencing groups
Berkeley	http://chlamydia-www.berkeley.edu:4231/
Genome Therapeutics	http://www.genomecorp.com/home.htl
Sanger	http://www.sanger.ac.uk/Projects/
Stanford	http://sequence-www.stanford.edu/group/ malaria/index.html
TIGR	http://www.tigr.org/tdb/mdb/mdb.html
University of Oklahoma	http://dna1.chem.uoknor.edu/index.html
University of Queensland	http://www.cmcb.uq.edu.au/aeruginosa/
University of Washington	http://chimera.biotech.washington.edu/ uwgc/
Washington University	http://genome.wustl.edu/gsc/bacterial/ salmonella.html
Tools and resources	
Biomolecular Tools	http://www.public.iastate.edu/;pedro/ rt_1.html
COGs	http://www.ncbi.nlm.nih.gov/COG/
NCGR	http://www.ncgr.org/microbe/Index_bns.html
MAGPIE	http://www.mcs.anl.gov/home/gaasterl/ genomes.html
Genobase	http://specter.dcrt.nih.gov:8004/
Micro Underground	http://www.lsumc.edu/campus/micr/mirror/ public_html/index.html
ANMR	http://www.wdcm.riken.go.jp/
WHO	http://www.who.ch/Welcome.html

Database or organization	Internet address
Pallen	http://www.qmw.ac.uk/;rhbm001/ketbook/ chapter.html
CDC	http://www.cdc.gov/
University of Kansas	http://www.kumc.edu/research/fgsc/main.html
University of Georgia	http://fungus.genetics.uga.edu:5080/
Tripos	http://www.tripos.com/sites.html
Motif	http://dna.Stanford.EDU/identify/
Pedant	http://pedant.mips.biochem.mpg.de/
GenTHREADE	http://globin.bio.warwick.ac.uk/genome/ genomic.html
MUMmer	http://sourceforge.net/projects/mummer/
Sybil	http://sourceforge.net/projects/sybil/
VISTA	http://genome.lbl.gov/vista/index.shtml
Prosite	http://ca.expasy.org/tools/scanprosite/
Interpro (families)	http://www.ebi.ac.uk/InterProScan/
Swiss-Prot	http://au.expasy.org/sprot/
TmHMM (transmembrane domain)	http://www.cbs.dtu.dk/services/TMHMM/

The conserve domain can be identified in the predicted protein. NCBI conserve domain search allows finding conserve domain in the proteins. The conserve domains are structurally independent units in the proteins. The analysis of conserve domain reveals the biological significance of the conserve domain in the evolutionary path of the proteins. The processes of domain assortment and gene duplication produce families of proteins in organisms. Following speciation, a newly derived genome will inherit the families of ancestor organisms, but will also develop new ones to meet evolutionary challenges. Comparison of each of the proteins encoded by an organism with every protein, an all-against-all comparison, reveals that which protein families have been amplified and what rearrangements have occurred during stages in the evolutionary process. [8]. Proteins shares the domains as the result of gene duplication events which generates variation in the sequences and also due to the structural rearrangement of domain which occurred during the evolutionary history of the protein.

Different proteins are mosaics of domains that occur in different combinations in a given protein. In a particular organism or group of organisms, one particular domain can be expanded to perform a particular function. Comparison of the domain content of an entire proteome with that of another proteome can reveal the biological roles of diverse domains in different organisms. Each of the predicted protein can further be used as query sequence for the database search against the proteome of other organisms [9]. The significant alignment gives the list of protein which can be identified as orthologous proteins.

Drug Target Identification

The pathogenic microbial genomes can be used to identify and validate the novel drug targets for drug designing. The predicted protein can be used to know the corresponding metabolic pathways in the genome. The comparison of metabolic pathways of host and pathogen can be used to identify and validate the novel drug targets. Analysis of complete genomes allows us to compile a list of potential gene products and identify the functions. The genes which are essential for the survival of pathogen but absent in host (humans) can be identified by performing comparative metabolic pathway analysis using metabolic pathway databases like Kyoto Encyclopedia of Genes and Genomes. This reduces the problem of searching for potential drug targets from a large list to selecting from a chosen few. As most currently known, antibacterial compounds are essentially inhibitors of certain bacterial enzymes, all enzymes specific to bacteria can be considered as potential drug targets.

Genome Comparison

The comparison of two or more completed genome of one organism can be carried out to find the degree of conservation and differences in the genomic features, this will give new area of research known as comparative genomics. The Comparative genomics includes a comparison of gene number, gene content, and gene location in both prokaryotic and eukaryotic groups of organisms [10].

The availability of complete genome sequences makes possible a comparison of all of the proteins encoded by one genome, the proteome of that organism, with those of another. Because the genome sequence provides both the sequence and the map location of each gene, both the sequence and location can be compared If a set of genes is grouped together at a particular chromosomal location, and if a set of similar genes is also grouped together in the genome of another organism, these groups share an evolutionary history

[11] (Table 1). Different types of database search tools like BLAST (basic local alignment search tool), multiple sequence alignment tool clustalw, VISTA can be used for genome comparison of different species.

EST Databases Search

The complete genomes of many organism is not available, instead the large collection of EST sequences obtained by random sequencing of cDNA copies of cell mRNA sequences is present in the biological database. The EST libraries are useful for preliminary identification of genes by database similarity searches. A more detailed analysis may then be made by cloning and sequencing the intact cDNA [12]. An EST database of an organism can be analyzed for the presence of gene families, orthologs, and paralogs. A protein from the yeast or fly proteome, for example, can be used as a query of a human EST database by translating each EST sequence in all six possible reading frames. The program TBLASTN is used for EST database search (Figure 2).

Genome Annotation

Once we succeed in identifying the gene, gene products, similar sequences and various biological features of newly sequenced DNA, the next major task is to annotate the sequence with generated information. Various Computational tools such as MAGPIE and GENEQUIZ can be used for annotation purpose.

The ultimate step in genome analysis is to collect the information found on gene and proteins sequences, alignments, gene function and location, protein families and domains, relationship of genes to those in other organisms, chromosomal rearrangements, and so on, into a comprehensive database.

Genome database may also be interfaced with other types of data. This database should be logically organized so that all types of information are readily accessible and easily retrievable by users who have widely divergent knowledge of the organism [13]. This goal is best achieved by using controlled vocabulary facilitate the search for novel relationships that can identify the same genetic or biochemical function in different organisms. The complete information generated about the genome under investigation after the genome analysis and annotated sequences and their web links to a variety of related information will be available to biological community (Table 1).

In summary, the availability of genome sequences provides an unprecedented opportunity to explore genetic variability both between organisms and within the individual organism. After the genes have been identified, we can use the outcome of genome analysis where great deal of information about the genome sequence such as the location and structure of genes (exons and introns), links to related sequences, detailed information about the function, biochemistry and structure of proteins and other gene products, literature links, links to genetic maps, location of sequence repeats including transposable elements, the location of sequence tag sites and SNPs will be available for further applications.

Conclusion

While genome analysis has already proved its worth over and over, we are at the beginning of its flowering. To a large extent, the results obtained to date through genome analysis have been the low-hanging fruit.

As methods get refined and some of the foundational research in models and algorithms gets incorporated into Bioinformatics tools, more difficult targets will be reached.

References

[1] Adams M.D., Celniker S.E., Holt R.A., Evans C.A., Gocayne J.D., Amanatides P.G., Scherer S.E., Li P.W., Hoskins R.A., Galle R.F., et al. (2000). The genome sequence of *Drosophila melanogaster. Science 287,* 2185–2195.

[2] Andrade M.A., Brown N.P., Leroy C., Hoersch S., de Daruvar A., Reich C., Franchini A., Tamames J., Valencia A., Ouzounis C., and Sander C. (1999). Automated genome sequence analysis and annotation. *Bioinformatics, 15,* 391–412.

[3] Bailey L.C., Jr., Fischer S., Schug J., Crabtree J., Gibson M., and Overton G..C. (1998). GAIA: Framework annotation of genomic sequence. *Genome Research, 8,* 234–250.

[4] Baker P.G., Goble C.A., Bechhofer S., Paton N.W., Stevens R., and Brass A. (1999). An ontology for bioinformatics applications. *Bioinformatics 15,* 510–520.

[5] Bansal A.K. (1999). An automated comparative analysis of 17 complete microbial genomes. *Bioinformatics 15,* 900–908.

[6] Bennetzen J.L., SanMiguel P., Chen M., Tikhonov A., Francki M., andAvramova Z. (1998). Grass genomes. *Proceedings of National Academy of Sciences, USA, 95,* 1975–1978.

[7] Bernardi G. (1995). The human genome: Organization and evolutionary history. *Annual Review of Genetics, 29,* 445–476.

[8] Blattner F.R., Plunkett G., III, Bloch C.A., Perna N.T., Burland V., Riley M., Collado-Vides J., Glasner J.D., Rode C.K., Mayhew G.F., Gregor J., Davis N.W., Kirkpatrick H.A., Goeden M.A., Rose D.J., Mau B., and Shao Y. (1997). The complete genome sequence of *Escherichia coli K-12. Science 277,* 1453–1462.

[9] Brown M.P., Grundy W.N., Lin D., Cristianini N., Sugnet C.W., Furey T.S., Ares M., Jr., and Haussler D. (2000). Knowledge-based analysis of microarray gene expression data by using support vector machines. *Proceedings of National Academy of Sciences, USA, 97,* 262–267.

[10] Bult C.J., White O., Olsen G.J., Zhou L., Fleischmann R.D., Sutton G.G., Blake J.A., FitzGerald L.M., Cawthon R.M., O'Connell P., Buchberg A.M., Viskochil D., Weiss R.B., Culver M., Stevens J., Jenkins N.A., Copeland N.G.. and White R.(1990). Identification and characterization of transcripts from the neurofibromatosis 1 region: The sequence and genomic structure of EVI2 and mapping of other transcripts. *Genomics 7,* 555–565.

[11] Cherry J.M., Ball C., Weng S., Juvik G., Schmidt R., Adler C., Dunn B., Dwight S., Riles L., Mortimer R.K. and Botstein D. (1997). Genetic and physical maps of *Saccharomyces cerevisiae. Nature (suppl. 6632) 387,* 67–73.

[12] Chervitz S.A., Aravind L., Sherlock G., Ball C.A., Koonin E.V., Dwight S.S., Harris M.A., Dolinski K., Mohr S., Smith T., Weng S., Cherry J.M., and Botstein D. (1998). Comparison of the complete protein sets of worm and yeast: Orthology and divergence. *Science 282,* 2022–2028.

[13] Bruijn F.J., Lupski J.R., and Weinstock G.M., Eds. 1998. Bacterial genomes: Physical structure and analysis. *Chapman and Hall, New York.*

In: Advances in Biotechnology
Editors: H. K. Dhingra, P. N. Jha et al.
ISBN: 978-1-62417-842-9

Chapter 2

COUNTERACTING MICROBES

Uma S. Dubey [*] ***and Anyaa Mittal***
Department of Biological Sciences,
Birla Institute of Technology & Sciences, Pilani
Pilani, Rajasthan, India

ABSRACT

Human survival in an adverse environment that contains diverse microbes depends on its ability to counter their pathogenic ability. The body is able to protect itself by virtue of various components of the innate immune system, acquired immune system and their interactions at various levels. Pathogens are only able to enter the body by breaking through the first line of defense mechanisms. The entry is followed by the body's mounting of a non specific innate immune response and finally the highly specific acquired immune response that even possesses 'immunological memory'.

This chapter covers the following topics of the non specific or the innate immune system- (i) Anatomical and Physiological barriers (ii) Inflammation and Phagocytosis (iii) Cell mediated effectors like Natural killer cell and Antibody dependent cellular cytotoxicity (iv) Receptor mediated elimination responses (v) Counteracting chemicals and (vi) Complement system mediated attack. It further moves on to describe the highly specific acquired immune system and the intricate interaction between its various subcomponents. Lastly, it deals with the body's

[*] Corresponding Author: Uma S. Dubey, Assistant Professor. Email: uma@bits-pilani.ac.in.

immune response to four major classes of microbes, namely, viruses, bacteria, fungi and protozoal parasites.

INTRODUCTION

We live in an environment which is replete with a large number and diversity of microorganisms. Many of these are very pathogenic and are a cause of serious public health concern. Fortunately, our immune system is endowed with a constellation of mechanisms to counteract these microbes. Coexistence in the same environment has been the driving force behind evolution of the microbes, as well as, of the immune system itself. Emergence of newly born infectious diseases like Acquired Immunodefficiency Syndrome (AIDS), Severe Acute Respiratory Syndrome (SARS), Dengue, etc. are the evidence for the existence of such an interplay at the microbial level. The ability of our immune system to counter any possible pathogens by generation of equally phenomenal diversity of T cells, B cells as well as MHC repertoire provide further evidence for existence of such a co-evolution. The survival, as well as, disease remittance of patients infected by pathogens, depends upon the interaction between the host system and the microbe at various levels of complexity. These conditions may be altered by therapeutic intervention.

The immune system constitutes structures and processes that protect the organism against disease by eliminating pathogens and malignancies. It detects a wide variety of agents, complex foreign compounds to viruses and further to parasitic worms, and needs to distinguish them from the organism's own healthy cells and tissues in order to function properly. Functionally, an immune response can be divided into two related activities—recognition and response. The immune system is able to recognize subtle chemical differences that distinguish one foreign pathogen from another. Furthermore, this system is able to discriminate between foreign molecules and the body's own cells and proteins. Once a foreign organism has been recognized, the immune system recruits a variety of cells and molecules to mount an appropriate response, called an effector response, to eliminate or neutralize the organism. Later exposure to the same foreign organism induces a 'immunological memory', which is characterized by a more rapid and heightened immune response that serves to eliminate the pathogen and prevent recurrence of the disease that is caused by the same infective agent [1].

The immune system may respond in either a non specific manner as in the natural or innate immune system or in an antigen specific manner as in adaptive or acquired immune system. The less specific component, innate immunity, provides the first line of defense against infectious agents. Most components of innate immunity are present before the onset of infection and constitute a set of disease-resistance mechanisms that are not specific to a particular pathogen but that include cellular and molecular components that recognize classes of molecules peculiar to frequently encountered pathogens. Structural barriers such as skin; phagocytic cells, such as macrophages and neutrophils; and a variety of antimicrobial compounds synthesized by the host; all play an important role in innate immunity. In contrast to the broad reactivity of the innate immune system, which is uniform in all members of a species, the specific component, namely, the acquired immune response does not come into play until there is an antigenic challenge to the organism. Acquired immunity responds to the challenge with a high degree of specialization and is characterized by memory [2]. Most disease conditions result from complex interaction between the mechanisms of pathogenicity of microbe and the multiple mechanisms of resistance of host. This interation results in either removal of infection or progression of disease.

In the present chapter we will discuss the innate as well as the acquired immune responses against microbes. We will also elucidate how our body deals with bacterial, viral, fungal and protozoal infectious agents.

Innate Immunity against Microbes

Microbial invasion in our body leads to generation of numerous effector cells and molecules. A localized elimination response is generated by inflammation. Inflammation leads to migration of phagocytic cells to the site of infection where they phagocytose and kill microbes. But before all this begins, the entry of microbe is prevented by certain structural features.

Anatomical and Physiological Barriers

Our entire body is covered by either skin or mucosal membrane which provides a formidable barrier against all microbes. Mucosal lining covers the respiratory, gastrointestinal, urinogenital as well as ocular surfaces. Presence of ciliated epithelium further helps diminish infectivity of microbes gaining access through the respiratory surface. There is an increased risk of infections in patients whose ciliated epithelial surface has to be bypassed in favor of respiration via tube during tracheostomy.

Normal homeostatic condition within the body at various levels itself can become inhibitory for the proliferation of a broad class of microbes. For example, body temperature only permits a very narrow range of microbes to survive within. Some of these like intestinal bacteria are symbiotic while others are normally commensals. Moreover, the elevated temperatures achieved during fever are detrimental to microbes on one hand, and have an augmenting effect on the immune system on the other. Both, cellular and humoral immune response can be influenced by body temperature. Endogenous pyrogens like IL-1 (α & β), IL-6 and TNF-α are products of macrophages. The binding of these cytokines with their receptors activate the arachidonic acid pathway. This pathway releases PGE2, the ultimate mediator of fever [3]. A beneficial role is indicated by an increase in motility, microbicidal activity and proliferative response of lymphocytes at elevated temperatures. The production as well as activity of cytokines like IL-1, IL-2 and interferons (IFNs) increases at higher temperatures [4].

INFLAMMATION AND PHAGOCYTOSIS

Microbe induced tissue injury leads to generation of vasodilatory factors (eg histamine) and chemotactic molecules like IL-8. This results in an increase in diameter of blood vessels (vasodilation), and migration of cells to the site of infection (chemotaxis) as directed by chemokines, complement component's breakdown products and N-formyl peptides produced by breakdown of bacterial proteins. Movement of cells from within the blood vessels to the outside intercellular fluid (extravasation) is facilitated by interaction between mucins and integrins present on cell surface of migrating cells and selectins and intracellular adhesion molecules (ICAM) present on the endothelial walls of the blood vessels. Leucocytes reaching the tissue bind to extracellular matrix via integrins and CD44. Acute phase proteins produced by the liver are also recruited to the site of infection by proinflammatory cytokines like IL-8 and TNF-α. In the process of phagocytosis microbes are endocytosed, degraded and remnants are exuded. Lysosomal degradation takes place enzymatically, either by reactive oxygen or by nitrogen containing intermediates. Pain is induced at the site of infection by virtue of prostaglandins [5].

CELL MEDIATED ELIMINATION MECHANISMS

Natural Killer Cells

NK cells have the inherent ability to kill virally infected cells and cancer cells. Morphologically they appear as larger lymphocytes with numerous cytoplasmic granules and are therefore sometimes called large granular lymphocytes (LGLs). They express multiple cell surface markers like CD2, CD56, LFA-1 and ICAM-1. They also have receptors for INF-γ, IL-12 and receptor for Fc of immunoglobulin, which increases their capacity of cell lysis.

NK cells distinguish healthy cells from cancer cells or virally infected cells through a balance between activating signals and inhibiting signals upon binding with respective ligands. These cells recognize their targets by multiple receptors which fall under two categories – the Killer Activatory Receptors (KARs) and the Killer Inhibitory Receptors (KIRs). The KARs delivers an activating signal whereas the KIRs deliver inhibitory signals upon binding with respective ligands. Because signals from inhibitory receptors have veto power over activating receptors, a negative signal from any inhibitory receptors can block lysis of targets by NK cells. The inhibitory receptor recognizes MHC I. The reduced expression of this by virally infected cells or cancer cells leads to their killing by NK cells. Thus, cells expressing normal levels of unaltered MHC I molecules tend to escape all forms of NK mediated killing. The cytolytic ability is mediated by perforins and granzymes released by NK cells. The perforin monomers polymerize to form transmembrane channels which punch a hole in the cell membrane of the target cells allowing the passage of granzymes into it. This leads to osmotic imbalance in the targets and their subsequent death [1, 5-6].

Antibody Dependent Cellular Cytotoxicity

It is the killing of antibody coated target cells in which effector cells bind the Fc region of the antibody and lyse the target cell in a non phagocytic manner. The effectors of Antibody Dependent Cellular Cytotoxicity (ADCC) are NK cells, macrophages, monocytes, neutrophils and eosinophils. The specificity of these cells towards antigens is attained by antibody molecules that coat these target cells. These cells possess receptors for antibodies which interlink them to the antigens. Multiple different mechanisms may be involved in target cell killing. The cell damage done by eosinophils and NK cells involve the release of perforin from these cells, while the binding of macrophages and eosinophils to the targets result in release of lytic enzymes at the site of Fc receptor binding. Moreover NK cells and macrophages secrete Tumor Necrosis Factor (TNF) after antibody links them to the target cells. The interaction between TNF and its receptor results in cytolysis [2].

RECEPTOR MEDIATED ELIMINATION RESPONSES

Elimination of target cells is facilitated by the linkage of target specific receptors to the effector cells. The receptors of innate and adaptive immune response differ from each other in characteristics related to specificity. The acquired immune response possess highly specific receptor whereas the adaptive immune response have broad based specific receptors against a wide range of microbes and are called Pattern Recognition Receptors (PRRs). The PRRs recognize structural moieties commonly shared amongst diverse microbes. Pattern Recognition Receptors are important in the innate immune response. They may exist either in soluble or cell associated form. Cell associated PRRs have the ability to recognize broad class of molecules commonly present on a wide range of microbes. For example, a form of cell associated PRRs; the Toll Like Receptors (TLRs) cause transcriptional activation, synthesis and secretion of cytokines which promote inflammatory response. These lead macrophages and neutrophills to the site of infection by chemotaxis. The recognition site of PRRs in microbes is called Pathogen Associated Molecular Patterns (PAMP). TLRs have been assigned numbers 1 to 11 (TLR1-TLR11). Interaction of TLRs with their specific PAMP induces NF-κB signaling and MAP kinase pathway and therefore the secretion of pro-inflammatory cytokines and costimulatory molecules released following TLR activation signal to other cells of the immune system, making TLRs key elements of innate immunity and adaptive immunity [7].

The soluble PRRs bind to free microbes to promote their opsonization and complement mediated lysis in body fluids. The initiators of complement system like Mannose binding lectin (MBL) and C-reactive proteins (CRP) are examples of soluble pattern recognition receptors. PRRs are perfectly able to distinguish between self and non self molecules because their ligands are absent in mammalian host cells.

COUNTERACTING CHEMICALS

Many counteracting proteins, like hydrolytic enzymes and cytokines are present in blood, intracellular fluids as well as various body secretions. Lysozymes are hydrolytic enzymes present in all mucosal secretions, such as, tears, saliva and nasal mucus. They lyse Gram positive bacteria by cleaving its peptidoglycan cell wall. Defensins are small cystein rich cationic proteins active against bacteria, fungi and viruses. They form permeable channels through the microbial cell membrane. TNF is a cytokine which kills microbes by either forming ion permeable channels or sending programmed cell death (PCD) inducing signals. IFN-α and IFN-β induce the production of antiviral proteins which contribute to the ability of the host to resist or recover from a viral infection. IFN-γ can help modulate immune responses by up regulating expression of Class I and Class II MHC molecules on target cells.

COMPLEMENT SYSTEM MEDIATED ATTACK

Complement system is a set of at least 20 proteins present in the blood. The activation of this system eventually results in the disruption of cell membranes and destruction of invading microorganisms. This occurs via the formation of membrane attack complex (MAC) which punctures multiple holes into the target, leading to osmotic imbalance and death. This system can be initiated by the classical pathway that depends on antigen-antibody complex; the mannose dependent lectin pathway; or the alternate pathway that is initiated by foreign surfaces. In all three pathways, a C3-convertase cleaves and activates component C3, creating C3a and C3b, and causing a cascade of further cleavage and activation events. C3b binds to the surface of pathogens, leading to greater internalization by phagocytic cells by opsonization. C5a is an important chemotactic protein, helping recruit inflammatory cells. C3a is the precusor of an important cytokine (adipokine) named ASP (acylation-stimulating protein) and is usually rapidly cleaved by carboxypeptidase B. Both C3a and C5a have anaphylatoxic activity, directly triggering degranulation of mast cells as well as increasing vascular permeability and smooth muscles contraction. MAC is the cytolytic endproduct of the complement cascade. It consists of C5b, C6, C7, C8 and polymeric C9 it forms a transmembrane channel, which causes osmotic lysis of the target cell.

Complement system breakdown products also mediate inflammation, chemotaxis, opsonization and cell lysis [8].

Acquired Immune Response

The adaptive or acquired immune response comes into play after foreign microbes are exposed to the antigen presenting cells (APCs). Antigens entering our body are phagocytosed by APCs and than subsequently processed by degradation within the lysosomes (antigen processing) and expressed on the surface of the cells in association with MHC (antigen presentation). It has two components: the cell mediated immune (CMI) response and the humoral immune (HI) response. The cell mediated immune responses is mediated by APCs (macrophages and dendritic cells), B cells and two types of T cells, namely, Cytotoxic T cell (T_C) and the Helper T cell (T_H). B cells mediate the humoral immune response of the body by secretion of antibodies. The cytotoxic T cells act either by destroying the pathogens directly or by killing the intracellularly infected cells. The helper T cells which constitute the other class of T cells regulates the immune response either by secretions of cytokines or by contact mediated signaling with T cells, B cells or APCs.

Cell lysis mediated by CTLs takes place by inducing apoptosis via two mechanisms: (i) Ca^{2+}-dependent exocytosis of cytolytic granules containing perforin and granzyme B; (ii) Ca^{2+}-independent cell death triggered by release or upregulation of CD95L and CD95 engagement on the target cell. In the target cells there is an activation of a specific class of cysteine proteases, the caspases, which execute apoptosis by cleaving a number of cellular proteins culminating in the structured disassembly of a cell [9, 10].

Interaction of a lymphocyte with its specific target leads to activation followed by intracellular cell signaling. The down steam effect of lymphocyte activation is its proliferation and subsequent differentiation into effector cells. A lyphocyte specific protein tyrosine kinase (lck) is associated with cytoplasmic tail of CD4 and CD8 coreceptors of T_H and T_C cells respectively. Activation involves phosphorylation of residues of T_CR and CD3. A phosphorylation cascade involving ZAP-70, LAT (linker of activated T cells) P13k and phoshplipase C (PLC) is initiated. This cascade terminates in mobilization of Ca^{2+} and activation of transcription factors such as NFAT, NF-κB and AP-I. These factors regulate cytokines like IL-2 that promotes long term proliferation and differentiation of the lymphocytes [11].

Immune Response against Virus

Viruses are obligate intracellular parasites that replicate within the cell using the host's intracellular machinery. When a virus replicates in a host cell it disrupts normal cellular machinery and ultimately kills the host cell. This is the cytopathic effect of the virus. Moreover, the antiviral immune response that follows also damages the host cell along with the virus. Such an immune response that is itself harmful to the host cell is its immunopathological response.

If a cell possesses a virus specific receptor only then does the virus infect it. For example, HIV infects the T helper cells and macrophages because these cells express CD4, a cell surface protein that also acts as a receptor of the AIDS virus. Similarly, the Epstein Bar Virus (EBV) infects B cells by binding to their CR2 receptor. If the host cells lack the receptor for a virus then it will inherently be resistant to that virus, for example, mice cells lack receptors for polio viruses and thus are resistant to polio. It is for this reason that humans are inherently resistant to plant and many animal viruses. The host will also be resistant to a virus if it lacks some component necessary for the replication of that virus.

Resistance against viral infection and its recovery depends upon the interaction that occurs between virus and immune response of the host. The defenses mounted by the host may act directly on the virus or indirectly by killing the infected cell. Non-specific host defenses act early in the encounter and prevent or limit infection while the specific host defenses function act later on. The host defense mechanisms involved in a particular viral infection will vary depending on the type of virus, dose and portal of entry. For example, orthomyxo- and paramyxo-virus families infect the host cells by binding to sialic acid receptors. Sialic acid containing glycoproteins in mucus can thus compete for the cell receptors and diminish or prevent binding of virus to the cells.

Innate Immune Response against Virus

The early or innate immune responses against virus include interferons, natural killer cells, and macrophages. Type I Interferons (IFN α/β) function to inhibit viral replication in both infected and uninfected cells by inducing an antiviral state. They bind to their specific receptor to induce intracellular signaling terminating in activation of 30 or more gene products inducing antiviral action. Binding of Type I IFN and Type II IFNs (IFN- γ) to NK cells activate them to kill virally infected cells. NK cells and macrophages kill virally infected cells by ADCC via Fc portion of antibodies [12, 13]. Depletion of NK cells *in vivo* by administration of anti-NK cell antibodies leads to enhanced susceptibility to infections by viruses such as by murine cytomegalovirus and recombinant vaccinia virus.

Humoral Immune Response against Virus

Humoral immunity plays an important role in limiting the spread and re-infection of viral particles. Antibodies may act by several mechanisms like, binding to cellular receptor thereby preventing its entry into the viral cell (for example binding of influenza virus to sialic acid residues in cell membrane), agglutination of viral particles to reduce their number thus spread and making them more susceptible to phagocytosis. Secretary IgA neutralizes viruses entering through the mucosal routes.

Complement Mediated Response against Virus

Complement factors neutralize virus by several different mechanisms like by (i) blanketing the antibody coated virus rendering it incapable of binding to its receptor on the host cell; (ii) inducing clumping and thus reducing the effective cell number (iii) inducing opsonization such as binding of C3b on virus-antibody complex making it more susceptible to phagocytosis; (iv) binding antibody coated virus that posses lipid envelope and induce lysis; (v) inactivating certain viruses in absence of specific antibodies, for examples certain retroviruses which themselves can act as receptor for the subunit of C1.

T Cell Mediated Antiviral Mechanisms

Once the virus has gained entry inside the host cell, antibodies become ineffective and cell mediated immune mechanisms come into play to eradicate the infection. Cytotoxic T (T_c) cells and T_H1 cells are the main effectors of T cell mediated antiviral defences. Cytotoxic T cells recognize viral antigens present on infected cell surface along with MHC I and lyse these infected cells by the release and action of perforins. They can also kill virally infected cell by inducing apoptosis. T_H1 cytokines like IFN-γ and TNF-α, produced by the helper T cells enhance the cell mediated responses against virally infected cells.

Virus Induced Immunopathology

The host has been bestowed with a variety of defense mechanisms to protect against viral infections, but sometimes it is the immune response to the infection that is the direct cause of tissue injury. For example, infants infected with cytomegalovirus have circulating immune complexes that deposit in the kidney and joints resulting in pathology such as arthritis and glomerular nephritis. Another example is fatal hemorrhagic shock syndrome associated with dengue virus infection. In this instance, fixation of complement by circulating immune complexes results in release of products of the complement cascade leading to sudden increased vascular permeability, shock and death.

Many viruses are able to suppress the immune response and thereby overcome or minimize host defenses. The best example is HIV which infects the CD4+ cells thereby destroying the specific immune system [14]. Other viruses (*e.g.* measles virus) can also infect lymphocytes and affect their replication and differentiation.

Primarily, HIV infects and kills CD4+ T lymphocytes, which function as regulators and amplifiers of the immune response. In the absence of effective anti-retroviral therapy, the hallmark decrease in CD4+ T lymphocytes during AIDS results in a weakened immune system, impairing the body's ability to fight infections or certain cancers leading to death. The major mechanism for CD4+ T cell depletion is programmed cell death. Apoptotic cells are eradicated by two different mechanisms: either a Fas-mediated mechanism during activation-induced cell death (AICD) or as a result of HIV proteins (Tat, gp120, Nef, Vpu) released from infected cells stimulating apoptosis in uninfected bystander cells. There is also evidence that as AIDS progresses cytokine dysregulation occurs, and the overproduction of type-2 cytokines (IL-4, IL-10) increases susceptibility to AICD whereas type-1 cytokines (IL-12, IFN-γ) may be protective [15].

IMMUNE RESPONSE AGAINST BACTERIA

Bacteria enter the body by invading the first line of host defense mechanism. Depending on the number of organisms entering and their virulence, different levels of host defense are elicited. If the inoculum size and the virulence are both low, then localized tissue phagocytes may be able to eliminate the bacteria with an innate, nonspecific defense. Larger inoculums or organisms with greater virulence tend to induce an adaptive, specific immune response. Immunity to bacterial infections is achieved by means of antibody unless the bacterium is capable of intracellular growth, in which case delayed-type hypersensitivity plays an important role. When the host encounters any invading bacteria, the primary response against it is immediate and non-specific in form of inflammation and phagocytosis. A secondary specific response follows the initial response. This includes the antibody-mediated immunity and the cell-mediated immunity. The antibody-mediated mechanism is principal immune response against extracellular bacteria while cell-mediated immunity is the major defense and protective response against intracellular bacteria [16].

Multiple responses are elicited during bacterial invasion. The peptidoglycan in the cell wall of the Gram negative bacteria activates the compliment system generating C3b which leads to opsonisation of bacteria and enhances phagocytosis. Some bacteria produce endotoxins such as LPS, which stimulate the production of cytokines such as TNF-α, IL-1, and IL-6 by macrophages or endothelial cells. These cytokines can activate macrophages. Phagocytosis of bacteria by macrophages and other phagocytic cells is another highly effective line of innate defense. However, some types of bacteria that commonly grow intracellularly have developed mechanisms that allow them to resist degradation within the phagocyte. Moreover, bacteria present in food, face resistance by the host due to the presence of gut microflora which can competitively inhibit the binding of pathogens.

TLRs induce inflammatory response against bacteria through NF-κB, a cytoplasmic factor controlling transcription. It controls many cytokine genes like IL-1, IL-2, TNF, IFN, IL-8 and IL-21. Secreted PRPPs like complement system also play an important role in inflammation, opsonization, chemotaxis and lymphocyte activation against the microbe.

Immune Response against Extracellular Bacteria

Extracellular bacteria can be pathogenic either by inducing a localized inflammatory response or by producing toxins. The immune response against bacteria targets on eliminating the bacteria and neutralizing the effects of toxins released by the pathogen. Infection by extracellular bacteria induces production of humoral antibodies secreted by plasma cells in regional lymph nodes and the submucosa of the respiratory, urinogenital tracts and gastrointestinal tracts. Polysaccharides present on the bacterial surface induce the production of a strong primary as well as secondary immune response in the host via production of IgM and IgG respectively. These antibodies against the surface antigens and the toxins stimulate multiple different types of effector responses such as activation of the complement system, opsonization or enhance phagocytosis and neutralization the bacterial toxin [1].

Complement activation can lead directly to lysis of Gram negative bacteria. Antibody binds to accessible bacterial cell surface antigens along with the C3b component of complement system, increasing phagocytosis and hence removal of the bacterium from the host. Antibody mediated activation of the complement system can also induce localized production of immune effector molecules that help to develop an amplified and more effective inflammatory response. For example, the complement split products C3a, C4a, and C5a act as anaphylatoxins that induce local mast-cell degranulation and thus vasodilation and the extravasation of lymphocytes and neutrophils from the blood into tissue space. Other types of complement split products act as chemotactic factors for neutrophils and macrophages, which contribute to the buildup of phagocytic cells at the site of infection. For removal of bacterial toxin, the antibody against it binds to the toxin and neutralizes it, the antibody-toxin complexes are then cleared by phagocytic cells in the same way as any other antigen-antibody complex.

Immune Response against Intracellular Bacteria

Intracellular bacteria can survive and even replicate within even the phagocytic cells where they are inaccessible to antibodies. For eliminating intracellular bacteria, the host response is completely different from the response to extracellular bacteria. While innate immunity is not very effective against intracellular bacterial pathogens, intracellular bacteria can activate NK cells, either directly or indirectly by stimulating macrophages production of

IL-12 which is a powerful NK cell – activating cytokine. These NK cells provide an early defense against intracellular bacteria. Intracellular bacterial infections tend to induce a cell-mediated immune response, specifically, delayed type hypersensitivity. In this response, cytokines secreted by CD4+ T cells notably IFN-γ, which activates macrophages to kill phagocytosed pathogens more effectively [6].

Immune Response against Fungi

There are several mechanisms used by host immune system to counter fungal infections. The principal mediators of innate immunity against fungi are the neutrophils. These neutrophils release fungicidal chemicals such as lysozymal enzymes and reactive oxygen species (ROS). Fungi are also phagocytosed by neutrophils for intracellular killing. Other mechanisms include responses at the level of membrane structure, their receptors and humoral factors. The host defense mechanism varies with the type of fungal infection. For example, macrophages are involved in killing fungal cells in case of infection by *Cryptococcus*, whereas neutrophils are the main effector cells in killing *Candida albicans* and *Aspergillus fumigates*. Furthermore, the defense mechanism is largely dependent on the anatomical site of infection in human body. For example, yeast and spores can be easily phagocytosed but larger hyphae cannot be effectively digested in the same way. Besides this the host response may also depend on the morphology of the pathogenic fungi [17].

Fungal infection leads to the production of chemotactic factors, which help in recruitment of leukocytes to the site of infection. These chemotactic factors include chemokines, peptides derived from activation of complement system pathway and products synthesized by fungi. Chemokines regulate several immunological activities in addition to chemotaxis which are important in host response to fungal infections. These include angiogenesis, hematopoesis, cytokine induction, T_H cell differentiation and antigen presentation.

Amongst the adaptive immune response, cell-mediated immunity plays an important role in providing defense against the fungal infection. It has been reported that granulomatus inflammation is important for controlling fungal infection in a tissue. The induced cell-mediated immunity depends on the fungal infection. The response can be categorized into T_H1-type CMI and T_H2-type CMI. In general, T_H1-type CMI is protective, required for clearance of a fungal infection. T_H1 cells produce INF-gamma type cytokines; promote cell-mediated immunity and phagocyte activation. The T_H2 responses are antibody mediated and sometimes detrimental to host [18]. This form of immunity leads to susceptibility to infection or allergic response. It tends to promote antibody production through the action of cytokines like IL-3 and IL-4.

Humoral immunity is mediated by production of certain protective antibodies. But this is effective only when these antibodies are produced in sufficient quantities. Functions of antibodies against fungal infections include prevention of adherence, toxin neutralization, antibody opsonization and ADCC.

Fungal infections represent a serious threat, particularly in immunocompromised patients. Interleukin-1β (IL-1β) is a key pro-inflammatory factor in innate antifungal immunity. Two signals are generally required for IL-1β production: an NF-κB-dependent signal that induces the synthesis of pro-IL-1β (p35), and a second signal that triggers proteolytic pro-IL-1β processing to produce bioactive IL-1β (p17) via caspase-1-containing multiprotein complexes called inflammasomes. It has been demostrated that tyrosine kinase Syk, operating downstream of several immunoreceptor tyrosine-based activation motif (ITAM)-coupled fungal pattern recognition receptors, controls both pro-IL-1β synthesis and inflammasome activation after cell stimulation with *Candida albicans*. Nlrp3 (also known as NALP3) was identified as the critical NOD-like receptor family member that transduces the fungal recognition signal to the inflammasome adaptor Asc (Pycard) for Caspase-1 (Casp1) activation and pro-IL-1β processing *in vivo* [19].

Aspergillus conidia and hyphae induce nuclear factor (NF)-κB translocation, and release of TNF-α and MIP-2 in a TLR2- and TLR4-dependent manner [17]. Neutrophil recruitment is severely impaired in mice lacking both functional TLR2 and TLR4, but is less impaired in single TLR2 or TLR4 deficient mice, suggesting that both receptors are required for an optimal immune response to Aspergillus [20]. The finding that innate and adaptive immunity to common fungal pathogens is mediated by the coordinated action of members of the TLR family acting through MyD88 raises the possibility that TLR manipulation may be a useful modality for induction of antifungal resistance [21].

Immune Response against Protozoal Parasites

Parasites are organisms that depend on a host either totally or partially for meeting their nutritional requirements and/or for completion of their life cycle. Protozoa vary greatly in their structural and biochemical properties and stimulate distinct patterns of immune responses. They are known to have evolved unique mechanisms for evading specific immunity. Protozoa activate quite distinct specific immune responses, which are different from the responses to fungi, bacteria and viruses. Some protozoa are human parasites, causing diseases. Examples of human diseases caused by protozoa include malaria, Leishmaniasis, Amoebiasis, etc. When pathogenic protozoa enter the body they often survive and replicate by adapting to the resisting natural host defenses. Protozoa cause chronic and persistent infections because they have evolved multiple mechanisms for evading and resisting specific immunity of the host.

Protozoa may be phagocytozed by macrophages, but many are resistant to phagocytic killing and may even replicate within macrophages. *Trypanosoma brucei*, the protozoa associated with sleeping sickness, is the best example of protozoa which can induce humoral immune response because of its extra-cellular location. In Leishmaniasis cellular defense mechanisms depend upon CD4+ T-lymphocytes and activated macrophages as effector cells that are regulated by cytokines of T_H1 subset. Malaria's causative agent, *Plasmodium* induces diversity of defense mechanisms which can be either cellular or humoral, depending on its location and stage of life cycle.

Conclusion

Our body has evolved highly intricate and efficient immune mechanisms to counter the pathogens. The microbes too have their own strategies of evading the immune system. The interaction of immune system with infectious organisms is a dynamic interplay of host mechanisms aimed at eliminating infections and microbial strategies designed to promote their own survival. This results in a very high degree of competition between the two interacting species and ultimately leads to co-evolution of both.

REFERENCES

[1] Thomas, J. K., Goldsby, R. A., Osborne, B.A. & Kuby, J. (2006) *Kuby Immunology*. 6th edition.

[2] Walter, F. (2003). *Medical Physiology: A Cellular and Molecular Approaoch*. Elsevier/Saunders.

[3] Boulant, J.A. (1991*). Thermoregulation. In Fever: Basic mechanisms and management.* Raven Press, New York.

[4] Sompayrac, L., (2003). *How the Immune System Works*, *Blackwell Publishing.*

[5] Abbas, A.K.. & Lichtman, A. (2003). *Cellular and Molecular Immunology*. Saunders Copyright.

[6] Doyle, S. L. & O'Neill, L.A. (2006). Toll-like receptors: From the discovery of NF-κB to new insights into transcriptional regulations in innate immunity. *Biochem. Pharmacol.* 72, 9, 1102–1113.

[7] Goldman, A.S. & Prabhakar, B.S. (1996). *The Complement System.* In: *Baron's Medical Microbiology*. Univ of Texas. Medical Branch.

[8] Lamkanfi, M., Festjens, N., Declercq, W., & Berghe, T.V. (2007). Caspases in cell survival, proliferation and differentiation. *Cell Death and Differentiation* 14, 1, 44–55.

[9] Feig, C. & Marcus, E. P. (2007). How apoptosis got the immune system in shape. *Eur. J. Immunol.* 37, 61-70.

[10] Mustelin, T. & Taskén, K. (2003). Positive and negative regulation of T-cell activation through kinases and phosphatases. *Biochem. J.* 371, 1, 15–27.

[11] Koyama, S. I. K. (2008). Innate immune response to viral infection. *Cytokine*. 43, 3, 336-341.

[12] Pier, G.B., Lyczak, J.B. & Wetzler, L. M. (2004). *Immunology, Infection, and Immunity*. ASM Press.

[13] Mogensen, T. E. (2010). Innate immune recognition and activation during HIV infection. *Retrovirology,* 7: 54.

[14] Alimonti, J. (2003). Mechanisms of CD 4+ T lymphocytic cell death in HIV infection and AIDS. *J. Gen. Virol.* 84,7,1649-1661.

[15] Berche, P. (2003). Bacterial aggression. *Ann Pharm Fr.* 61, 4, 270-275.

[16] Shoham, S. (2005). The immune response to fungal infections. *British Journal of Haematology.*129, 569-682.

[17] Blanco, J. & Gracia, M.E. (2008). Immune response to fungal infections. *Veterinary Immunology and Immunopathology*.

[18] Gross, O., Poeck, H, Bscheider, M., Dostert, C., Hannesschlager, N., Endres, S., Hartmann, G., Tardivel, A., Schweighoffer, E., Tybulewicz, V., Mocosai, A., Tscopp, J. & Ruland. J. (2009). Syk kinase signalling couple to the Nlrp3 inflammasome for antifungal host defence. *Nature.* 459, 433-436.

[19] Meier, A., Kirschning, C.J., Nikolaus, T., Wagner, H., Heesemann, J. & Ebel, F. (2003) Toll-like receptor (TLR) 2 and TLR4 are essential for Aspergillus-induced activation of murine macrophages. *Cell Microbiology,* 5, 561–570.

[20] Bellocchio, S., Montagnoli, C., Bozza, S., Gaziano, R., Rossi, G.,Mambula, S.S., Vecchi, A., Mantovani, A., Levitz, S.M. & Romani, L.(2004). The contribution of the Toll-like/IL-1 receptor superfamily to innate and adaptive immunity to fungal pathogens *in vivo. Journal of Immunology*, 172, 3059–3069.

In: Advances in Biotechnology ISBN: 978-1-62417-842-9
Editors: H. K. Dhingra, P. N. Jha et al.

Chapter 3

POLYMERIC NANOPARTICLES: A NOVEL CARRIER FOR DRUG DELIVERY

Vishnu Agarwal[1,*] and Himanshu Pandey[2]
[1]Department of Biotechnology,
Motilal Nehru National Institute of Technology, Allahabad, India
[2]Department of Pharmaceutical Sciences, Faculty of Health & Medical Sciences, Sam Higginbottom Institute of Agriculture, Technology & Sciences, Allahabad, India

ABSTRACT

Nanotechnology can be defined as the science and engineering involved in the design, synthesis, characterization and application of materials and devices whose smallest functional organization in at least one dimension is on the nanometer scale (one-billionth of a meter) [1, 2]. In the past few years nanotechnology has grown by leaps and bounds, and this multidisciplinary scientific field is undergoing explosive development [3-6]. It can prove to be a boon for human health care, because nanoscience and nanotechnologies have a huge potential to bring benefits in areas as diverse as drug development, water decontamination, information and communication technologies, and the production of stronger, lighter materials. Human health-care nanotechnology research can definitely result in immense health benefits. The genesis of

* Corresponding Author. E-mail: vishnua@mnnit.ac.in, himanshu_leo29@rediffmail.com.

nanotechnology can be traced to the promise of revolutionary advances across medicine, communications, genomics, and robotics [7, 8].

INTRODUCTION

The prefix 'nano' derives from the Greek word for 'dwarf'. One nanometer (nm) is equal to one-billionth of a meter, or about the width of 6 carbon atoms or 10 water molecules. A human hair is approximately 80,000 nm wide, and a red blood cell is approximately 7000 nm wide. Atoms are smaller than 1 nm, whereas many molecules including some proteins range between 1 nm and larger [9].

The conceptual underpinnings of nanotechnologies were first laid out in 1959 by the physicist Richard Feynman in his lecture, 'There's plenty of room at the bottom'. Feynman explored the possibility of manipulating material at the scale of individual atoms and molecules, imagining the whole of the Encyclopedia Britannica written on the head of a pin and foreseeing the increasing ability to examine and control matter at the nanoscale. The term 'nanotechnology' was not used until 1974, when Norio Taniguchi, a researcher at the University of Tokyo, used it to refer to the ability to engineer materials precisely at the nanometer level. The primary driving force for miniaturization at that time came from the electronics industry, which aimed to develop tools to create smaller (and therefore faster and more complex) electronic devices on silicon chips. Furthermore, at IBM in the United States, a technique called electron beam lithography was used to create nanostructures and devices as small as 40 to 70 nm in the early 1970s.

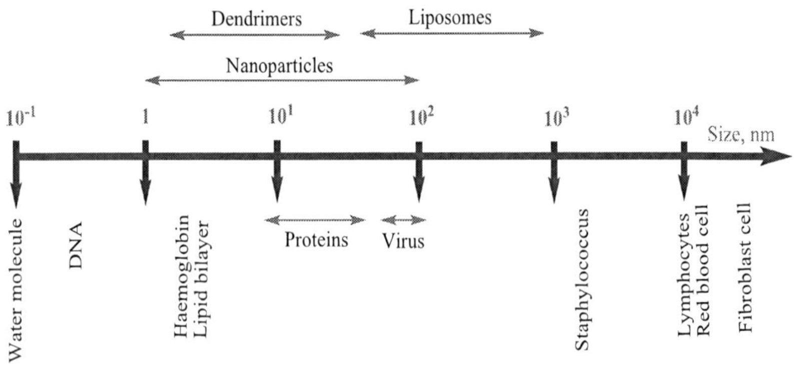

Figure 1. The size spectrum showing nanoparticles and other nanosystems.

NANOTECHNOLOGY IN DRUG DELIVERY

From nanotechnology there is only one step to nanomedicine, which may be defined as the monitoring, repair, construction, and control of human biological systems at the molecular level, using engineered nanodevices and nanostructures [10]. It can also be regarded as another implementation of nanotechnology in the field of medical sciences and diagnostics. One of the most important issues is the proper distribution of drugs and other therapeutic agents within the patient's body [11].

During the past two decades, however, researchers involved in the development of pharmaceuticals have understood that drug delivery is a fundamental part of drug development, and a wide range of drug delivery systems has thus been designed. Ideally, all these systems would improve the stability, absorption, and therapeutic concentration of the drug within the target tissue, as well as permit reproducible and long-term release of the drug at the target site [12,13]. In addition to reducing the frequency of drug administration and thus improving patient comfort, novel drug delivery systems would offer protection and improve the pharmacokinetics of easily degradable peptides and proteins, which often have short half-lives in vivo [14]. For the pharmaceutical industry the field of drug delivery represents a strategic tool for expanding drug markets, because new delivery technologies could repackage classical drugs, offering a competitive edge after the expiry of patents and avoiding competition from generics. Demonstrating this advantage clearly, 13% of the current global pharmaceutical market is related to the sale of products that include a drug delivery system [15].

The final aim of pharmaceutical research is the delivery of any drug at the right time in a safe and reproducible manner to a specific target at the required level [16]. For many drugs, however, these ideal requirements constitute hype rather than hope. For example, although the oral route is one of the preferred methods of drug delivery, because it is noninvasive, adequate peptide or protein drug delivery has not yet been attained via this route [17]. This is partly due to the acidic conditions of the stomach, the first-pass effect of the liver (i.e., the loss of drug as a result of metabolic processes that occur before it enters the systemic circulation), and the resistance exerted by the intestine—all of which alter, destroy, or reduce absorption of nearly all macromolecules, thus reducing their bioavailability. As a result, millions of diabetics worldwide have to self administered insulin injections daily, provoking a high percentage of negligence in this treatment.

Drug bioavailability is a related problem with potential nanotechnology solutions. Nanotechnology is opening new therapeutic opportunities for many agents that cannot be used effectively as conventional oral formulations because of their poor bioavailability. In some cases, reformulation of a drug with smaller particle size may improve oral bioavailability [18]. Nanoparticles formulations provide protection for agents susceptible to degradation or denaturation in regions of harsh pH, and also prolong the duration of exposure of a drug by increasing retention of the formulation through bioadhesion [19].

Polymeric Nanoparticles

Polymeric nanoparticles (Fig.2) provide an additional advantage over non-polymeric nanoparticles viz. carbon nanotube, silica nanoparticles, etc due to some inherent properties like biocompatibility, nonimmunogenicity, nontoxicity and biodegradability. These are colloidal carrier, 10 nm -1μm in size, consisting of synthetic or natural polymers. Polymeric nanoparticles are a broad class comprised of both vesicular systems (nanocapsules) and matrix systems (nanospheres). Nanocapsules are systems in which the drug is confined to a cavity surrounded by unique polymeric membrane whereas nanospheres are systems in which the drug is dispersed through out the polymer matrix. The various natural polymers like gelatin, albumin and alginate are used to prepare the nanoparticles; however they have some inherent disadvantages like poor batch-to-batch reproducibility, prone to degradation and potential antigenicity [20].

Figure 2. Scanning electron microscopy image of polymer nanoparticles.

Polymeric nanoparticles have many advantages such as:

- Easy to fabricate and characterize
- Process is inexpensive so the final cost of formulation is reduced.
- Polymers are biocompatible, biodegradable, non toxic and non-immunogenic that are suitable for drug delivery application.
- Mostly polymers are water soluble such as chitosan and sodium alginate.
- Applicable to a broad category of small molecules (drugs), genes, proteins and polynucleotides.
- Can be lyophilized and are stable after reconstitution with medium for application.
- Do not produce irritation to body tissues.
- Mucoadhesive property of polymer (for example chitosan) increases the retention time at the site that is helpful for bioavailability improvement.

METHODS FOR PREPARATION OF POLYMERIC NANOPARTICLES

There are large numbers of methods (Table 1) to prepare polymeric nanoparticulate systems. These depend to a large extent on the nature of the polymer and the drug, to be delivered.

Table 1. Methods of Manufacturing polymeric nanoparticles

Method
Solvent displacement
Salting out
Emulsion diffusion
Emulsion- solvent evaporation
Coacervation
Reverse micellar methods
In situ polymerization

Solvent Displacement Methods

This method, shown in Scheme 1, are perhaps the simplest means of preparing polymeric nanoparticles, provided that the polymer and drug can be solubilized in ethanol, acetone, or methanol as water- miscible solvents. The polymer in the presence of the aqueous phase of appropriate stabilizers such as polyvinyl alcohol (PVA) or polysorbate 80, forms remarkably narrow distribution nanoparticles [21,22].

Figure 3 shows a SEM image of cefuroxime axetil nanoparticles prepared by this method. The cefuroxime axetil nanoparticles is most likely to be formed in the amorphous state on first manufacturer. After a period of time, cefuroxime axetil crystallizes and its oral absorption from the nanoparticles is greatly reduced.

Salting Out

A variant on this technique is the one illustrated in Scheme 2, where a polymer is dissolved in a water- miscible organic solvent, which is added to a stirred aqueous gel containing a salting- out agent and a stabilizer [24].

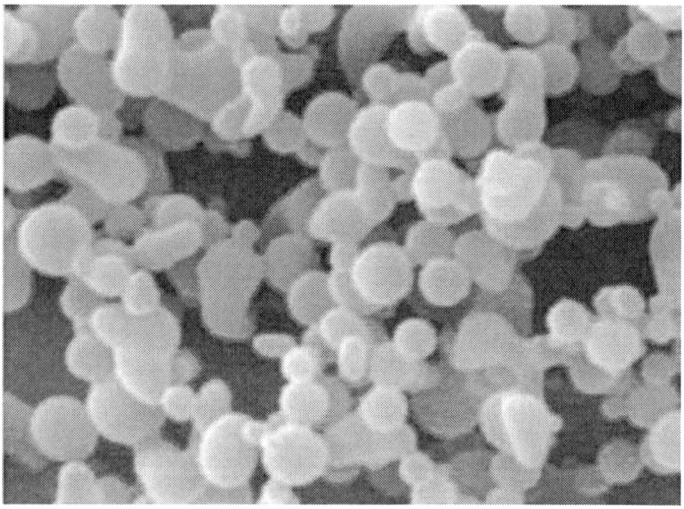

Figure 3. SEM image cefuroxime axetil nanoparticles (average diameter ~ 1oo nm), prepared by solvent evaporation method.

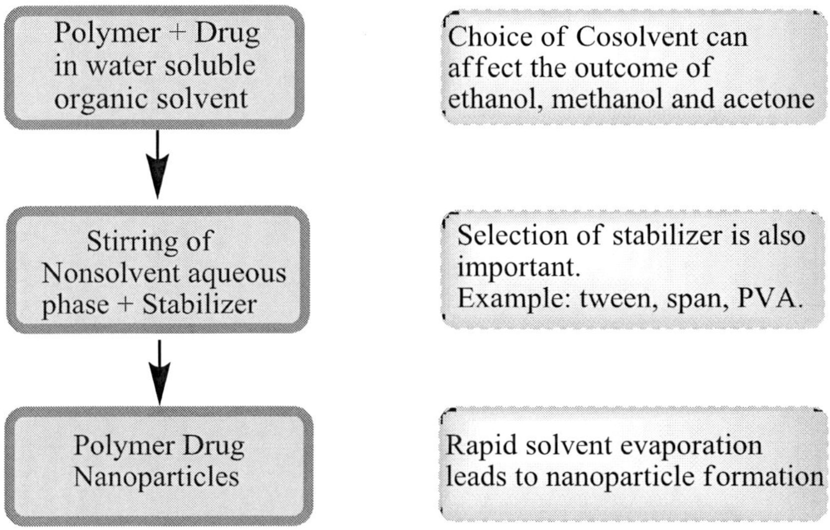

Scheme 1. A summary of the solvent- displacement method by which the particles in Figure 2 were prepared [23].

Emulsion Diffusion

The use of a partly water- miscible solvent in which the drug and polymer is dissolved is shown in Scheme 3. An emulsion is formed on addition of the polymer- drug solution to an aqueous phase containing a stabilizer. Because of the partial miscibility of the first solvent, solvent diffuses from the dispersed droplets into the bulk to provide after some time polymer nanoparticles containing drug [25,26].

Emulsion- Solvent Evaporation

Scheme 4 illustrate the different approach. It involves the dissolution of the drug- polymer mix in an organic solvent, which is immicible with water, so that on addition to an aqueous phase, an emulsion is formed in the presence of appropriate emulsifiers. On heating, the internal organic phase can be evaporated in a controlled fashion leaving the polymer- drug nanoparticles, whose size is dependent on the properties of the emulsion [27].

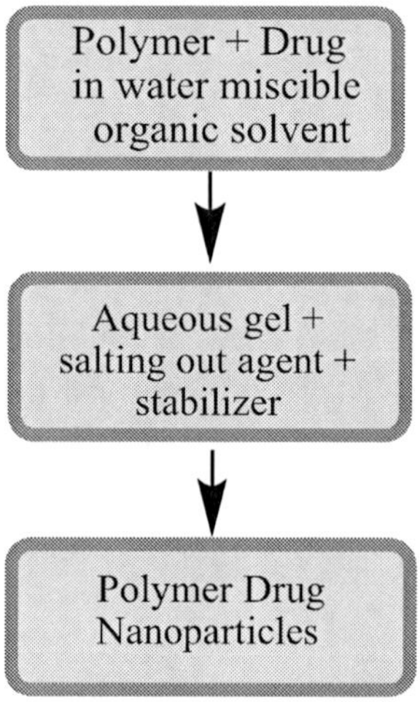

Scheme 2. Schematic of the salting out method in which a polymer and drug are dissolved in a water- miscible organic solvent efore addition to an aqueous gel with a salting- out agent and stabilizer.

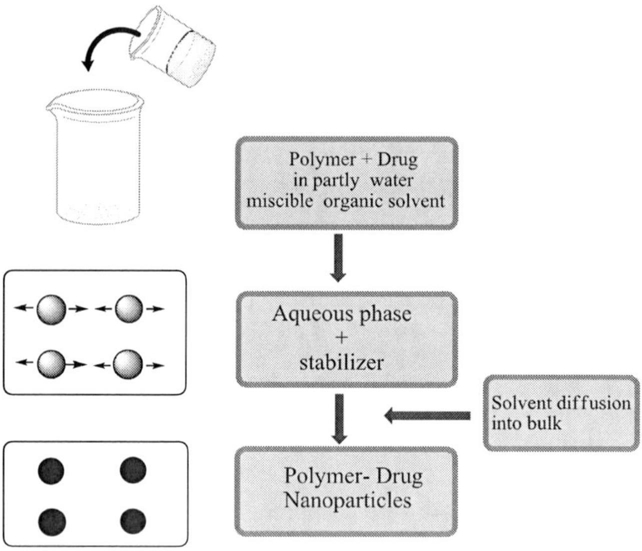

Scheme 3. Schematic of the emulsion technique where the drug and the polymer are dissolved in a partly water miscible solvent, which is then added to an aqoues phase where emulsion particles are formed. The solvent, being partly miscible in the water phase, diffuses from the droplets eventually leaving the nanoparticles as the droplets shrink.

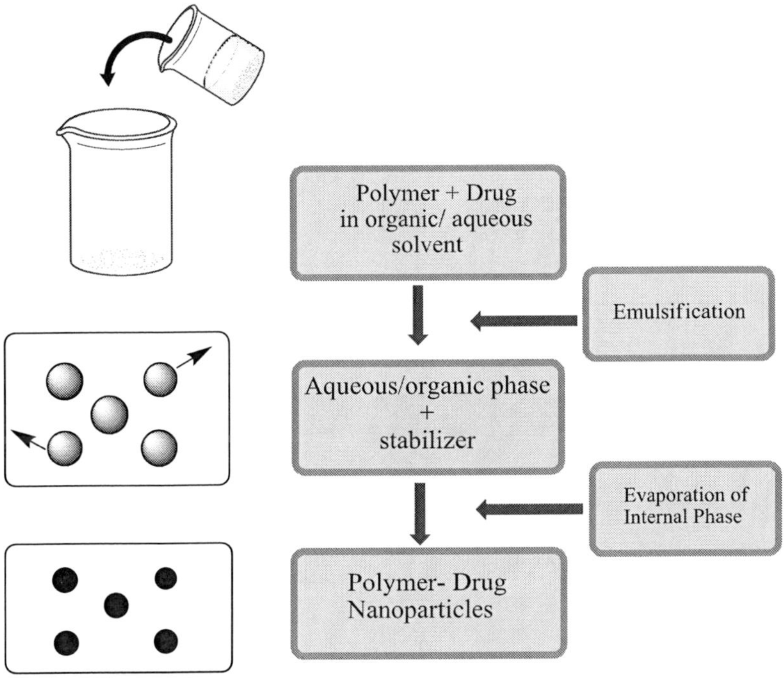

Scheme 4. Similar to the process in Scheme 3, where emulsification of the polymer/ drug organic solution takes place forming an oil-in- water system. The alternative is the dissolution of the polymer and the drug in an aqueous phase, which is added to an organic phase to form an water-in- oil emulsion. In both cases, the internal phase is then evaporated leaving the polymer- drug particles.

Reverse Micellar or Microemulsification Method

Figure 4 illustrates the formation of metallic nanoparticles using reverse microemulsions as an example of growing particles within individual dispersed phases or molecular containers. In this process, a metal salt is solubilized in the aqueous interior of microemulsion I. The aqueous phase of microemulsion II contains a reducing agent. When mixed, collision and coalescence of the droplets allow the chemical reaction to proceed, thus producing the precipitate of the metal or metal oxide within the confines of the droplet. Thus, droplet size determines to an extent nanoparticles size. Vesicles can be used for producing nanoparticles by acting as a template in a similar manner [28].

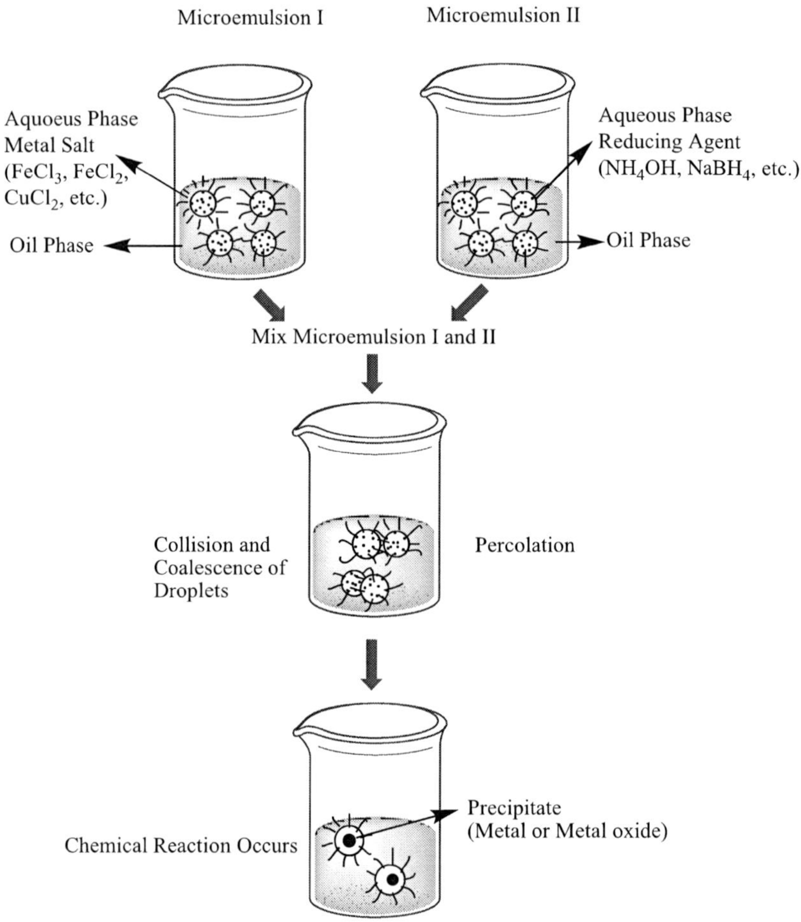

Figure 4. The preparation of nanoparticles by the inverse micelle process in which a chemical reaction between microemulsion or inverse micelles after collision and perhaps fusion converts the soluble salt into an insoluble metal or metal oxide as shown.

Coacervation

Coacervation has been a process used in the process used in the fabrication of microparticles for many decades since Bungenberg de Jong and Kruyt described the phenomenon ca. 1932. A coacervate is a gel- like liquid state of matter, which can be treated to produce solid microparticles or nanoparticles. There are several ways to form coacervates [29]. A modified coacervation technique is described by Calvo *et al.* [30] for the preparation of chitosan-alginate (CH–ALG) nanoparticles. Briefly, an aqueous solution of sodium alginate is sprayed onto the chitosan solution under continuous magnetic stirring at 1000 rpm for 30 min. Nanoparticles formed due to the interaction between the negative groups of ALG and the positively charged amino groups of CS (ionotropic gelation), were collected by centrifugation. For particle size and size distribution study these nanoparticles were re-dispersed in 5 ml of ultrapure water and studied by standard methods [31].

In Situ Polymerization

In this mode of synthesis of polymeric nanoparticles (e.g., of polybutylcycanoacrylates), monomeric materials (e.g., butylcyanoacrylate) are dispersed in a suitable solvent and emulsified. An initiator ig generally added to the system and the monomers condense to form a polymeric matrix, although in the case of the alkyl cyanoacrylates no initiator is required as the aqueous medium acts as the initiator of polymerization. As one example isobutylcyanoacrylate dissolved in ethanol can be mixed with an oil plus an oil- soluble drug to constitute one phase; the second phase is an aqueous solution of 0.5% Pluronic F68 (poloxamer 118). The two phases are mixed and the polyisobutylcyanocrylate is formed at the interface between the oil and the water [32].

Future Perspectives

There are likely to be multiple applications of nanotechnology in drug delivery. Nanotechnology can help in making nanodevices for complex surgical operation, like glaucoma, retinal vascular surgery and so on, and also in the development of new biocompatible polymeric material for transplantations. It will also benefit the different delivery formats: injectable, oral, implantable, transscleral, and so on. Nanotechnology formulations can also be used for resolving solubility issues and increasing the number of compounds available for potential development. Beside this, nanotechnology can also help to develop an effective and robust DNA nanoparticle therapy for the treatment of genetically based blinding diseases. It can even help in generation of scaffolds for tissue bioengineering, especially for neural stem cells and also use of these for the delivery of growth factors and the stem cells.

REFERENCES

[1] Emerich, DF. & Thanos, CG. (2003). Nanotechnology and medicine. *Expert Opin Biol Ther.* 3,655 - 63.

[2] Sahoo, SK.& Labhasetwar, V.(2003). Nanotech approaches to drug delivery and imaging. *Drug Discov Today,*8,1112- 20.

[3] Williams, D.(2004). Nanotechnology: a new look. *Med Device Technol,*15,9 - 10.

[4] Cheng, MM. Cuda, G. Bunimovich, YL. Gaspari, M. Heath, JR. & Hill, HD. et al.(2006) Nanotechnologies for biomolecular detection and medical diagnostics. *Curr Opin Chem Biol,* 10,11 - 9.

[5] Matsudai, M.& Hunt, G.(2005). Nanotechnology and public health. Nippon Koshu Eisei Zasshi,52,923- 7.

[6] Chan, WC.(2006). Bionanotechnology progress and advances. *Biol Blood Marrow Transplant,*12,87 - 91.

[7] Lammers, T. Subr V. Ulbrich K. Hennink W E. Storm G. & Kiessling F. (2010). Polymeric nanomedicines for image-guided drug delivery and tumor-targeted combination therapy. *Nano Today,* 5 (3),197-212.

[8] Mishra B. Bhavesh Patel B. & Tiwari S. (2010). Colloidal nanocarriers: a review on formulation technology, types and applications toward targeted drug delivery, *Nanomedicine: Nanotech Biol and Med,* 6(1),9-24.

[9] Whitesides, GM.(2003). The right size in nanobiotechnology. *Nat Biotechnol,* 21,1161- 5.

[10] Moghimi, SM. Hunter, AC.& Murray, JC.(2005). *Nanomedicine: current status and future prospects*. FASEB,19,311 – 30.

[11] Jain, KK.(2003). Nanodiagnostics: application of nanotechnology in molecular diagnostics. *Expert Rev Mol Diagn,* 3,153 - 61.

[12] Zhu Y. Che L. He H. Jia Y. Zhang J. & Li X. (2011). Highly efficient nanomedicines assembled via polymer–drug multiple interactions: Tissue-selective delivery carriers, *J Control Rel* (In Press, Uncorrected Proof, 22 March 2011).

[13] Kayser, O. Lemke, A. & Hernandez-Trejo, N.(2005). The impact of nanobiotechnology on the development of new drug delivery systems. *Curr Pharm Biotechnol,* 6,3 - 5.

[14] Orive, G. Hernandez, RM. Rodriguez Gascon, A. Dominguez-Gil, A.& Pedraz, JL. et al.(2003). Drug delivery in biotechnology: present and future. *Curr Opin Biotechnol,*14, 659- 64.

[15] Mazzola, L.(2003). Commercializing nanotechnology. *Nat Biotechnol,*21,1137- 43.

[16] Vasir, JK.& Labhasetwar, V.(2005). Targeted drug delivery in cancer therapy. *Technol Cancer Res Treat,*4,363- 74.

[17] Pawar, R. Ben-Ari, A.& Domb, AJ(2004). Protein and peptide parenteral controlled delivery. *Expert Opin Biol Ther,*4,1203- 12.

[18] Hu, L. Tang, X.& Cui, F.(2004). Solid lipid nanoparticles (SLNs) to improve oral bioavailability of poorly soluble drugs. *J Pharm Pharmacol,*56,1527- 35.

[19] Arangoa, MA. Campanero, MA. Renedo, MJ. Ponchel, G. & Irache, JM.(2001). Gliadin nanoparticles as carriers for the oral administration of lipophilic drugs. Relationships between bioadhesion and pharmacokinetics. *Pharm Res,*18,1521- 7.

[20] Nahar, M. Dutta, T. Murugesan, S. Asthana, A. Mishra, D. Rajkumar, V. Tare, M. Saraf, S.& Jain, NK.(2006). Functional polymeric nanoparticles: an efficient and promising tool for active delivery of bioactives. *Crit Rev Ther Drug Carrier Syst,* 23, 259-318.

[21] Fessi, H. Puisieux, F. Devissaguet, JP. Ammoury, N.& Benita, S.(1989). Nanocapsule formation by interfacial polymer deposition following solvent displacement. *Int J Pharm,*55,R1-4.

[22.] Govender, T. Stolnik, S. Garnett, MC. Illum, L.& Davis, SS.(1999). PLGA nanoparticles prepared by nanoprecipitation: drug loading and release studies of a water soluble drug. *J Control Release,*57,171-85.

[23] Peltonan, L.(2002). The effect of co-solvent on the formulation of nanoparticles. AAPS Pharm. *Sci. Tech,*3 (4), art 32.

[24] Toth, J. Kardos-Fodor, A.& Halasz-Peterfi, S.(2005). The formation of fine particles by salting-out precipitation. *Chem Eng Process,*44,193-200.

[25] Moinard-Checot, M. Chevalier, Y. Briancon, S. Beney, L.& Fessi, H.(2008). Mechanism of nanocapsules formation by the emulsion–diffusion process. *J Colloid Interface Sci,*317,458-68.

[26] Trotta, M. Debernardi, F.& Caputo, O.(2003). Preparation of solid lipid nanoparticles by a solvent emulsification–diffusion technique. *Int J Pharm,*257,153-60.

[27] Niwa, T. Takeuchi, H. Hino, T. Nohara, M.& Kawashima, Y.(1995). Biodegradable submicron carriers for peptide drugs: preparation of DL-lactide/glycolide copolymer (PLGA) nanospheres with nafarelin acetate by a novel emulsion-phase separation method in an oil system. *Int J Pharm,*121,45-54.

[28] Eastoe, J. Hollamby, MJ.& Hudson, L.(2006). Recent advances in nanoparticles synthesis with reversed micelles. *Adv Colloidal Interface Sci,*10,128- 130.

[29] Ramaswamy, C.& Florence, AT.(2005). Self-assembly of some amphipathic dendrons. *J Drug Sci Tech,*15, 307-311.

[30] P, Calvo. C, Remunan-Lopez. J.L, Vila-Jato. & M.J, Alonso.(1997). Novel hydrophilic chitosan–polyethylene oxide nanoparticles as protein carrier, *J. Appl. Polym. Sci.*, 63,125–132.

[31] S.K, Motwani. S, Chopra. S, Talegaonkar. K, Kohli. F.J, Ahmad. R.K, Khar. (2008). Chitosan– sodium alginate nanoparticles as submicroscopic reservoirs for ocular delivery: formulation, optimisation and in vitro characterization, *Eur. J. Pharm. Biopharm,* 68(3),513–525.

[32] Al-Khouri, N., Robolt-Truepel, L. Fessi, H.(1986). Development of a new process for the manufacture of polyisobutylcyanoacrylate Nanocapsules. *Int J Pharm,* 28,125-132.

In: Advances in Biotechnology
Editors: H. K. Dhingra, P. N. Jha et al. ISBN: 978-1-62417-842-9

Chapter 4

LIPASE CATALYZED TRANS-ESTERIFICATION REACTION FOR BIODIESEL PRODUCTION: CURRENT STATUS AND FUTURE PROSPECTS

Bhavya Ravi, Madhulika Rai, Sandhya Mehrotra and Rajesh Mehrotra*

Department of Biological Sciences,
Birla Institute of Technology and Science, Pilani, Rajasthan, India

ABSTRACT

Biodiesel fuels (BDFs) are essentially methyl or ethyl esters of fatty acids. The method for biodiesel production is called trans-esterification wherein triglycerides are converted into esters through reaction with simple alcohols. Lipases are hydrolytic enzymes that have become renowned for their ability to catalyze trans-esterification reaction. Although lipases are fairly ubiquitous enzymes synthesized by plants, animals and microbes, microbes have emerged as robust sources of the enzyme and efforts are being made to expand the available sources of lipase enzyme. The method of lipase catalyzed trans-esterification suffers from drawbacks such as high enzyme cost, low product yield, high

* Corresponding Author: Rajesh Mehrotra, E-mail:rajmeh25@hotmail.com.

reaction time and the need of organic solvents or water in the reaction mixture. In this chapter, we focus on advances that have been made over the years to improve several aspects of lipase catalyzed trans-esterification and to make the process of biodiesel fuel production cost effective, efficient and sustainable. Some of the strategies include the cloning of lipase genes and their subsequent expression in homologous and heterologous hosts to synthesize greater amounts of the enzyme, lipase immobilization on various porous and non-porous supports to improve enzyme stability and reusability, use of whole cell biocatalysts and novel cell surface display systems, bio-imprinting by which the active site of lipase can be frozen in a conformation most suitable for its enzymatic activity and protein engineering strategies such as Rational Protein Design (RPD) and directed evolution. All these strategies are aimed to improve the catalytic properties of lipase itself, reduce lipase costs and increased reusability of enzyme for several batches.

Keywords: Lipase catalyzed trans-esterification, biodiesel, microbial lipase, enzyme immobilization, protein engineering

INTRODUCTION

The two major problems confronting the world today are fossil fuel depletion and environmental degradation. The increasing demand for petroleum coupled with a steady decline in petroleum reserves has led to escalation of fuel costs and also of other petroleum derived products. For a sustainable future, the natural environment has to be protected. Greenhouse gas emissions (comprising chiefly of carbon di oxide, methane and nitric oxide) as a result of the use of fossil fuels has resulted in a steady rise in global temperatures and this global warming has brought along with it a plethora of problems.

The challenge now is to produce fuel and technology which is clean, eco-friendly and sustainable. Biodiesel, which is derived from plant oils and animal fats, has gained wide acceptance as a sustainable alternative to petroleum and diesel due to its environmental benefits and renewability. Biodiesel fuels (BDFs) are essentially methyl or ethyl esters of fatty acids. The method for biodiesel production is called trans-esterification wherein triglycerides- which are the main components of oils or fats are converted into esters through reaction with simple alcohols. The physical properties of biodiesel, like density and viscosity are very comparable to fuels derived from crude oil and therefore it can be used as it is or as an additive to currently used fuels without requiring much modification to conventional diesel engines.

LIPASE CATALYZED TRANS-ESTERIFICATION

Catalysis of trans-esterification reaction is broadly classified into two categories- chemical and enzymatic. Trans-esterification of oils and fats to make biodiesel can be represented by the following basic equation:

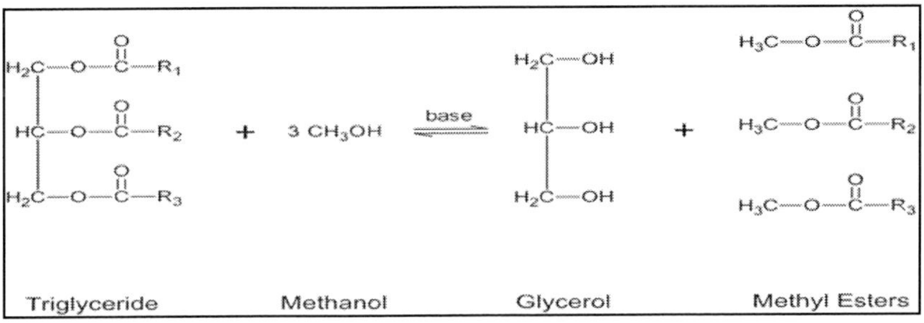

Figure 1. Trans-esterification Reaction.

After the reaction, the by-product glycerol is separated from the unreacted feed. Shah *et al.* [1] have demonstrated that a mixture of oil with excess of ethanol when refluxed at 70°C for 1 hour gave ethyl esters of fatty acids with a yield of 93%.

The disadvantage posed by acid/base catalyzed reaction are high energy consumption, difficulty in the trans-esterification of triglycerides with a high free fatty acid content and added cost of downstream processes like glycerol recovery, treatment of alkaline waste water, removal of inorganic salts and water from the product [1,2,3].

On the other hand, enzymatic catalysis allows for the synthesis of specific alkyl esters, easy recovery of glycerol and the trans-esterification of triglycerides with high free fatty acid content [1,4,5]. Lipases are hydrolytic enzymes (Triacylglycerol acylhydrolases E.C.3.1.1.3) which hydrolyze fats or triglycerides to free fatty acids and glycerol. These enzymes are however versatile and also catalyze a wide range of other reactions such as trans-esterification, aminolysis and acidolysis reactions making them important industrial biocatalysts [6]. Lipases have the capacity to act in micro-aqueous environment and catalyze esterification, trans-esterification, aminolysis and acidolysis [7]. Both fungal and bacterial lipases have been exploited by various industries [6].

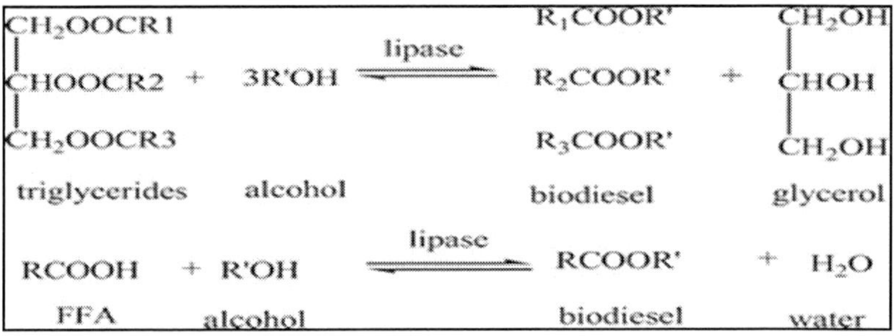

Figure 2. The two activities of lipase enzyme. Above: Hydrolytic Activity ; Below: Synthetic Activity.

Lipases have two important enzymatic activities:

1. Hydrolytic activity by which they catalyze the hydrolysis of triglycerides into free fatty acids
2. Synthetic activity by which they catalyze the synthesis of esters from glycerol and long chain fatty acids.

This synthetic activity of lipase is of main interest in BDF synthesis by lipase catalyzed trans-esterification. Lipases are now popular biocatalysts for biodiesel production through trans-esterification. However, even this method suffers from drawbacks such as high enzyme cost, low product yield, high reaction time and the need of organic solvents or water in the reaction mixture.

Advances have been made over the years by adopting novel strategies to improve several aspects of lipase catalyzed trans-esterification to make the process of BDF production cost effective, efficient and sustainable. In this chapter, we would be concerning all the recent advances in the field of lipase catalyzed biodiesel production.

Microbial Lipases

Lipases are ubiquitous enzymes that are widely distributed in plants, animals and microbes [8]. Microbial enzymes are often more useful than enzymes derived from plants or animals because of the great variety of catalytic activities available, the high yields possible, ease of genetic manipulation, regular supply due to absence of seasonal fluctuations and rapid growth of microorganisms on inexpensive media [6]. Microbial enzymes are also more stable than their corresponding plant and animal enzymes and their production is more convenient and safer [9]. Based on their synthetic activity, microbes can be screened and those that show a high synthetic activity are selected. To choose the suitable microorganisms showing good synthetic activity of lipases, they are screened by growing them on the suitable media-agar plate consisting of Oleic acid (free fatty acid), dodecanol and Rhodamine B. As dodecanol and oleic acid forms esters and water, in presence of lipase there are very less amount of free fatty acids available for Rhodamine B to bind with them and fluoresce under ultraviolet radiation. Therefore, the microorganisms producing more lipases having good synthetic activity are qualitatively chosen depending upon the intensity of fluorescence.

Table 1. List of some of the common microbes which are exploited for their lipase producing ability

Source	Reference
Candida rugosa	[1]
Candida albicans	[10]
Mucormeihei	[11]
Candida antarctica	[12]
Pseudomonas fluorescens	[13]
Rhizopusoryzae	[14]
Aspergillusniger	[15]
Rhizopuschinensis	[16]
Bacillus licheniformis	[17]
Bacillus subtilis	[18]
Bacillus cereus	[8]

Although enzymatic trans-esterification suffers from certain drawbacks, efforts are being made by scientists worldwide to overcome these limitations.

Lipase immobilization, whole cell biocatalyst system and molecular biology tools like cloning, protein engineering and directed evolution have shown promising results in the improvement of yield, enhanced enzyme activity and stability, reusability potential of enzymes for several batches, enhanced substrate specificity and enantioselectivity.

These have rendered the process more feasible and cost-effective. The following sections are devoted to the current advances that have been made in the abovementioned areas and their applicability in BDF production.

Lipase Immobilization

The process of lipase immobilization makes the overall process of biodiesel production economically feasible by facilitating the recovery of the catalyst for repeated use. Immobilization can perform additional functions such as improving the stability of the enzymes [19] in organic media and also enhance enzyme activity [20, 21]. There are several techniques that can be used for carrying out enzyme immobilization with each having its own merits and demerits. Adsorption onto solid supports is the most common approach. It is cheap, easily carried out, and tends to be less disruptive to the enzymatic protein than chemical means of attachment, the binding being mainly by hydrogen bonds, multiple salt linkages, and Van der Waal's forces.

The resulting immobilized system is not very stable because of the weak bonds involved and due to the desorption of the protein resulting from changes in temperature, pH and ionic strength. Another disadvantage is non-specificity further adsorption of foreign proteins or other substances to the support. Entrapment in gels is relatively inexpensive and is performed under milder conditions.

Covalent attachment and cross linking with bi-functional reagents produce very rigid bonds which could inactivate the catalytic site and destroy the enzymatic activity in the immobilization process. When trying to select the type of reaction by which a given protein should be immobilized, the choice is limited by the fact that the binding reaction must be performed under conditions that do not cause loss of enzymatic activity, and the active site of the enzyme must be unaffected by the reagents used.

There have been many trans-esterification processes that have used lipase in the immobilized form [1, 22-25]. Some of their different types, ME conversion percentages and reaction conditions have been summarized in Table 2. The two problems encountered when using immobilized systems are the loss of enzyme activity upon immobilization [26] and during recycling. The possible reasons for this are desorption, substrate inactivation or product inhibition [26, 27].

A method of pretreating lipase was developed by Lee *et al.* [28] to prevent loss of its activity during covalent immobilization. In this study, *Rhizopus oryzae* lipase was pretreated before immobilization and then immobilized on a silica gel surface. In this study, *Rhizopus oryzae* lipase was pretreated before immobilization and then immobilized on a silica gel surface.

Immobilized lipase pretreated with 0.1% of soybean oil had better activity than those pretreated with other materials (20 times higher than that of immobilized non-pretreated lipase) and its activity was maintained at levels exceeding 90% of its original activity even after 10 reuses.

Table 2. Immobilized lipase systems with yield of Methyl esters

Study	Immobilized System	Yield Of Methyl Esters
Chang *et al.* [29]	Novozym 435	97.9% (12.4 h)
Xu*et al.*[30]	Lipozyme TL IM	98% (12h)
Li *et al.* [31]	Novozym 435 combined with Lipozym TL IM	95% (12h)
Awang*et al.* [32]	*C. rugosa*on polyurethane foam	80%
Huang *et al.* [33]	Novozym435 and Lipozyme TL IM combined (RSM* applied)	97.2%(20h)
Shimada *et al.* [22]	*C. Antarctica* lipase	95% (48h)
Hsu *et al.* [34]	*P. cepacia* lipase on phyllosilicate sol-gel matrix	95% (18h)
Kumari*et al.* [35]	*E. aerogenes* immobilized on silica	94% (48h)

*RSM- Response Surface methodology: It is a powerful tool for the optimization of chemical reactions and industrial processes.

WHOLE CELL BIOCATALYSTS

In Whole-cell immobilization, unlike enzyme immobilization, the target cell producing the enzyme or expressing a specific enzyme on their surface is immobilized on the substrate. Immobilization of fungal mycelium within biomass support particles (BSPs) and expression of the lipase enzyme on the surface of yeast cells are two popular models of whole-cell biocatalyst systems that are designed to reduce enzyme associated process costs [36].

Entrapment techniques which encapsulate cells inside a matrix like polyurethane, alginate and collagen require harsh conditions resulting in a loss of cell viability. Whole cell immobilization technique uses the natural property of cells to adhere to any surface. Although this kind of adherence presents advantages as it is easy to perform, free of diffusion and does not affect the viability of cells, it is also very sensitive to changes in temperature, pH and ionic strength. Whole-cell immobilization techniques are cost effective as they reduce the cost of the process by avoiding the complex isolation, purification and immobilization of extracellular lipase [37]. Filamentous fungi have emerged as robust candidates for whole-cell biocatalysis [36].

Nakashima *et al.* [38] were able to demonstrate that cell aggregation that occurs as a result of immobilization acts as a trigger for the enhancement of intracellular lipase production by a *Rhizopus* species. Immobilization of *Rhizopus chinensis*cells on various BSPs made from polyurethane, nylon, polyester, polyvinyl alcohol and cellulose enhanced intracellular lipase as much as several fold than that of freely suspended cells irrespective of the BSP material.

In another study, to elucidate the reasons underlying the increased activity of whole-cell biocatalysts, Hama *et al.* [39] studied the localization of lipase in *R. oryzae* by western blotting analysis which showed that the species produced two lipases- a 34kDa lipase (ROL34) which was bound to the cell wall and a 31kDa lipase (ROL31) which was bound to the cell wall or membrane. In suspension cells, the amount of membrane bound lipase decreased sharply with cultivation time while in immobilized cells, large amounts of membrane bound lipase remained even in the later period of cultivation.

It was concluded that immobilization inhibits the secretion of ROL31 into the culture medium so it remains bound on the surface for long. Membrane localized ROL31 is chiefly responsible for the methanolysis activity of *R. oryzae* cells.

Novel *E coli* surface expression systems have been designed which either expresses certain proteins on their surface that facilitate and strengthen the adhesion properties or the enzyme responsible for the catalysis is itself expressed on the cell surface and remains bound for many reaction cycles increasing the efficiency of trans-esterification reaction.

Wang *et al.* [40] have investigated the specific adhesion of *E coli* with surface exposed cellulose binding domains (CBD). They studied the effect of pH, temperature and amount of surface expressed CBD to determine the optimum conditions for whole-cell immobilization. They generated *E coli* with surface expressed CBD enabling single-step immobilization onto different cellulosic supports. It was observed that cells with CBD bound specifically and strongly to cellulose supports at a wide range of pH. CBD showed highest binding to cellulose at 4°C, but highest cell loading was observed at 37°C and pH=7. Immobilized cells remained tightly bound to the cellulose support even after extensive washing.

It was concluded the affinity tags such as cellulose binding domains increase the specificity of interaction and strength of binding. Specific adhesion of whole cells to cellulosic materials with high affinity has been demonstrated on the surface of *E coli*. Recombinant cells that express CBD can bind rapidly to cellulose fibres, cellulose beads and sponges. This method represents a simple, mild and inexpensive technique of immobilization for the preparation of whole cell biocatalyst. Narita et al. [41] have developed a novel *E coli* cell surface display system by employing PgsA anchoring motif. Lipase B from *Candida antarctica* was displayed on *E coli* cell surface and showed an activity of 4.6 U/g (dry cell weight). The CALB (Lipase B) displaying *E coli* cells catalyzed enantio-selective trans-esterification and acted as effective whole cell biocatalyst systems. They succeeded in showing that through fusion to its C terminus; the PgsA protein can be used in *E coli* to display enzymes in a conformationally active form.

There have been many similar studies [41-46] aimed at the development of an *E coli* cell surface display systems for biotechnological and industrial applications. In a recent review by Wu *et al.* [47] the authors have discussed comprehensively many versatile surface display systems that have been developed for the purpose of bioremediation and biofuel production. Hashemizadeh *et al.* [48] have performed a comparative study of immobilized whole cell and commercial lipase as a catalyst for biodiesel production. *Rhizopusoryzae* whole cell biocatalyst immobilized on BSP was compared with Novozym 435 (the most effective extracellular immobilized lipase) for the methanolysis of soybean oil in a solvent free medium.

ME (Methyl esters) content in the reaction mixture was 84 wt% using *R. oryzae* whole cell biocatalyst under optimum conditions and at a reaction time of 48 hours which is comparable with the yield of 90 wt% obtained by Novozym 435.

Table 3 summarises results for ME yields obtained by various studies employing whole cell biocatalyst that have employed different immobilization systems and trans-esterification conditions.

Table 3. Comparative list of Whole cell biocatalysts

Study	Type of Whole Cell Biocatalyst	Yield of Methyl Esters
Matsumoto *et al.* [42]	Cell surface display system using FLO1 anchor	78.3% (72h)
Adachi *et al.* [45]	*A.oryzae*coexpressing triglyceride and partial glyceride lipases	98%
Ban *et al.*, 2001 [49]	*R. oryzae* immobilized on BSP	90%(72h)
Ali *et al.* [11]	*Pseudomonas* sp.immobilized on sodium alginate	70%(48h)
Chen *et al.* [50]	*R.oryzae* on porous nonwoven fabric in a packed bed reactor. RSM was applied.	72.6%(24h)
Xiao *et al.*[15]	Novel *A.Niger* strain immobilized on BSP	87%(72h)
Li *et al.* [51]	*R.oryzae* immobilized on BSP	90%(48h)
Xiao *et al.* [52]	Mold fungus *JN7* immobilized on BSP, microalgae lipid feedstock	68.2%(48h)
Pazouki*et al.* [53]	*R.oryzae* immobilized on BSP	98.4%(72h)
Ying *et al.*[54]	*B.subtilis* immobilized in hydrophobic magnetic polymicrosphere (Magnetic cell	90%(72h)

Genetic Engineering

Genetic and environmental manipulation may be employed to increase the yield of cells [55] and their enzyme activity by making the enzyme of interest constitutive, inducing it or by producing altered enzymes [56]. This process is fairly easy using microbial cells because of their short generation times, relatively simple nutritional requirements and relative ease of screening procedures for the desired characteristic.

Cloning and sequencing of lipase genes have been an area of attraction for a majority of researchers worldwide. Many bacterial lipases have been cloned, sequenced and expressed in homologous and heterologous hosts. The most widely used expression host is *E. coli* [57, 58], while other efficient hosts include *P. pastoris* and *S. cerevisiae* [59].

So far, several lipase genes from filamentous fungi, yeast and bacteria have been cloned and expressed. But compared with effective expression of lipase from other organisms, expression of *Pseudomonas* lipase in yeast is rarely reported. In a study, a lipase gene (*lipA)* of *P. fluoriscens* 26-2 was cloned and integrated into *P. pastoris* genome via some strategies to generate recombinants with Mut^+ (normally utilize methanol) and Mut^S (utilize methanol slowly) phenotypes. Lipase was methanol induced, expressed and after inducible expression was purified. It was observed that maximum activity of Mut^+ recombinants could reach 20.3 IU/ml after methanol inducible expression for 72 hours while Mut^S recombinant just reached 5.6 IU/ml. Also, Ca^{2+} could significantly enhance the activity of lipase upto 2.8 fold while EDTA could dramatically reduce it. The secreted expression of cloned lipase gene in *P. pastoris*settled a foundation to construct high level expression recombinants suitable for large scale preparation [60]. In yet another study, to enhance the production of *B. cepasia* G63 lipase, *lipA* and *lipB* genes were expressed by a high copy number vector pBBR1Tp in homologous strains.

The cloned lipase was purified and its properties were studied revealing that it had high temperature stability, organic solution tolerance and transesterification capacity for soyabean oil substrate [61]. Based on these characteristics, it can be used as a potential biocatalyst in biodiesel production. Luo *et al.* (2006) have demonstrated that a lipase gene *lipB68,* could be cloned and expressed in *E. coli* BL21. As a result a recombinant LipB68 protein was formed which produced biodiesel with a yield of 92% after 12 hrs., at the lowest temperature of 20°C and was suggested as a competitive energy saving biocatalyst.

LIPASE OPTIMIZATION

We have discussed many methods that aim to improve the process of lipase catalyzed biodiesel production and make the entire process cost effective and sustainable. Most of them have focused on identifying the most suitable biocatalyst or the best reaction conditions for efficient application. Approaches such as Lipase immobilization, whole cell biocatalysts, use of solvent systems to improve methanol endurance, surface expression of enzymes on whole cell biocatalysts, recombinant lipase producing yeast and powerful optimizing techniques such as Response surface methodology (RSM) have shown great promise in terms of improved conversion rates of FAs, enzyme recovery, lowering of enzyme cost and sustainability. There still remains one powerful tool, that of lipase engineering which aims at improving the catalytic efficiency of lipases as biocatalysts for biodiesel production. There is yet another approach termed ‘Bioimprinting’ aimed at improving the catalytic activity of lipase.

BIO-IMPRINTING

In this technique, some reversible changes are induced in the three dimensional structure of lipase by non-covalent interactions with a ligand using lyophilization and an anhydrous solvent [6]. The closing of the lipase lid is caused by water, and the opening is caused by an oil–water interface in oil–water two-phase system.

This phenomenon, called interfacial activation, has been found for *Candida rugosa* lipase (CRL) upon its adsorption onto hydrophobic supports [62], and especially onto polypropylene powder [6,63]. Both, closing and opening, cannot take place in organic solvents, where the lids of the lipase molecules will be closed because of the absence of an interface.

Bio-imprinting has been introduced as a technique of interfacial activation of lipase for anhydrous reaction applications. The immobilized lipase is fixed in an “open conformation” and enhanced enzymatic activity is achieved. In a novel approach by Mingarro *et al.* [64], it was attempted to generate "activated" lipolytic enzyme by molecular imprinting, which would exhibit an improved catalytic activity in nonaqueous media.

The rationale was to "trap" (freeze-dry) the enzyme in a presumably activated form when it is bound to an amphiphile-water interface and to further assay it after "washing out" the print amphiphile with an anhydrous solvent in water-restricted environments where its imprinted conformation is expected to be preserved. A dramatic enhancement in the activity of the enzyme was observed.

This approach was further verified by Gonzalez-Navarro *et al.* [65] who tested its applicability for several lipases where it was found that the method allows conversion yields after short reaction times which are remarkably higher than those obtained with the corresponding powdered commercial samples. Other studies [62, 65-67] in this field have furthered our understanding of how to use this technique effectively to enhance lipase characteristics to favor its role as a biocatlyst.

PROTEIN ENGINEERING

Bornscheuer *et al.* [68] emphasize on three key strategies that have been developed to expand the range of accessible and improved lipases. These strategies include

1. The exploration of biodiversity to expand the sources and number of new biocatalysts
2. The use of rational protein design to improve enzymes for which the 3-D structure has been elucidated or homology-modeled.
3. Directed evolution

Rational Protein design makes use of computer modeling to elucidate the basis for substrate specificity, fatty acid chain length specific and regio-/enantio selectivity. Directed evolution represents an efficient alternative to rational protein design when the structure and mechanism of the targeted enzymes are not available. In short, this method consists of the generation of large mutant libraries using various methods for random mutagenesis and recombination followed by advanced high-throughput screening or selection methods [69] to identify desired variants. The most prominent methods are error-prone PCR [70] and DNA shuffling [71]. Schmid and co-workers [72,73] investigated the molecular bases of stereoselectivity of two different lipases by SDM (Site Directed Mutagenesis) based on molecular modeling studies, and were able to generate mutants with altered stereoselectivity compared with that of the wild-type enzyme.

In yet another study, Liebton and coworkers demonstrated that the enantioselectivity of a lipase from *Pseudomonas aeruginosa* PAO1 towards 2-methyldecanoate could be enhanced by means of directed evolution from E = 1 to E ~26 for a mutant bearing five amino acid substitutions [74].

All the protein engineering studies discussed thus far have aimed at creating a line of mutant lipases which function as the best biocatalysts for a wide range of biotechnological and industrial applications.

Developments in this field when coalesced with improvements in the areas like reaction engineering and the design of efficient biocatalyst systems would result in the development of novel and optimized processes for BDF production. With these developments, we can hope to one day uncover the most eco-friendly, inexpensive, renewable and sustainable fuel to power our ideas in the future.

References

[1] Shah S., Sharma S. and Gupta M.N. (2004). Biodiesel Preparation by Lipase-Catalyzed Transesterification of *Jatropha* Oil. *Energy Fuels,* 18: 154-159.

[2] Du W., Xu Y.Y., Liu D.H. and Li Z.B. (2005). Study on acyl migration in immobilized lipozyme TL-catalyzed transesterification of soybean oil for biodiesel production. *Journal of Molecular Catalysis B: Enzymatic*37:1-6: 68-71.

[3] Al-Zuhair S. (2005). Production of Biodiesel by Lipase-Catalyzed Transesterification of Vegetable Oils: A Kinetics Study. Biotechnology Progress, 21:5:1442–1448.

[4] Nelson L.A., Foglia F.A. and Marmer W.N. (1996). Lipase-catalyzed production of biodiesel. *J. Am. Oil Chem. Soc.* 73:1191-1195.

[5] Kumari V., Shah S. and Gupta M.N. (2007). Preparation of Biodiesel by Lipase-Catalyzed Transesterification of High Free Fatty Acid Containing Oil from *Madhucaindica. Energy Fuels.*,21:1:368–372.

[6] Sangeetha R., Arulpandi I. and Geetha A. (2011). Bacterial Lipases as Potential Industrial Biocatalysts: An Overview. *Research Journal of Microbiology,* 6:1-24.

[7] Joseph B., Ramteke P.W., Thomas G. and Shrivastava N.(2008). Standard Review Cold-active microbial Lipases: A versatile tool for industrial applications. *Biotechnology and Molecular Biology Review*, 2:2:39-48.

[8] Dutta S. and Ray L. (2009). Production and Characterization of an Alkaline Thermostable Crude Lipase from an Isolated Strain of *Bacillus cereus* C_7. *Applied Biochemistry and Biotechnology.,* 159:1:142-154.

[9] Wiseman A.(1995). Handbook of enzyme biotechnology. 3rd ed. 3–8.

[10] Hube B., Stehr F., Bossenz M., Mazur A., Kretschmar M. and Schäfer W.(2000). Secreted lipases of *Candida albicans*: cloning, characterization and expression analysis of a new gene family with at least ten members. *Arch Microbiol,.* 174:362–374.

[11] Ali A., Kaur M. and Mehra U.(2011). Use of immobilized *Pseudomonas* sp. As whole cell catalyst for the trans-esterification of used cotton seed oil. *Journal of Oleo Science*, 60:1:7-10.

[12] Chen H., Ju H., Wu T., Liu Y., Lee C., Chang C., Chung Y. and Shieh C.(2011). Continuous Production of Lipase-Catalyzed Biodiesel in a Packed-Bed Reactor: Optimization and Enzyme Reuse Study. *Journal of Biomedicine and Biotechnology.* Article ID 950725.

[13] Luo Y., Zheng Y., Jiang Z., Ma Y. and Wei D.(2006) A novel psychrophilic lipase from *Pseudomonas fluorescens* with unique property in chiral resolution and biodiesel production via transesterification. *Appl. Microbiol. Biotechnol.,* 73:349-355.

[14] Kaieda M, Samukawa T, Matsumoto T, Ban K, Kondo A, Shimada Y, Noda H, Nomoto F, Ohtsuka K, Izumoto E and Fukuda H.(1999). Biodiesel fuel production from plant oil catalyzed by *Rhizopusoryzae* lipase in a water-containing system without an organic solvent. *Journal of Bioscience and Bioengineering,* 88:6:627-631.

[15] Xiao M., Qi C., Obbard J.P. (2011). Biodiesel production using *Aspergillusniger* as a wholecell biocatalyst in a packed-bed reactor. *GCB Bioenergy,* 3:293–298.

[16] Sun S.Y. and Xu Y.(2008). Solid-state fermentation for "whole-cell synthetic lipase" production from *Rhizopuschinensis*and identification of the functional enzyme. *Process Biochemistry*, 43:219–224.

[17] Sangeetha R., Geetha A. and Arulpandi A. (2010). Concomitant production of protease and lipase by *Bacillus licheniformis* VSG1: Production, purification and characterization. *Brazilian J. Microbiol.*, 41: 179-185.

[18] Ahmed E.H., Raghavendra T. and Madamwar D.(2010). An alkaline lipase from organic solvent tolerant *Acinetobacter* sp. EH28: Application for ethyl caprylate synthesis. *Bioresour. Technol.*, 101: 3628-3634.

[19] Lee D.G., Ponvel K.M., Kim M., Hwang S., Ahn I.S. and Lee C.H.(2009). Immobilization of lipase on hydrophobic nano-sized magnetite particles. *Journal of Molecular Catalysis B: Enzymatic.*, 57:62-66.

[20] Guncheva M., Zhiryakova D., Radchenkova N. and Kambourova M. (2009). Properties of immobilized lipase From *Bacillus stearothermophilus* MC7. Acidolysis of triolein with caprylic acid. *World Journal of Microbiology and Biotechnology*, 25:4:727-731.

[21] Liu C., Lin Y., Chen C. and Chang J. (2009). Characterization of *Burkholderia* lipase immobilized on celite carriers. *Journal of the Taiwan Institute of Chemical Engineers*, 40:4:359-363.

[22] Shimada Y., Watanabe Y., Samukawa T., Sugihara A., Noda H., Fukuda H. and Tominaga Y. (1999). Conversion of Vegetable Oil to Biodiesel Using Immobilized *Candida antarctica*lipase. *Journal of the American Chemical Society,* 76:7.

[23] Watanabe Y., Shimada Y., Sugihara A., Noda H., Fukuda H. and Tominaga Y. (2000). Continuous Production of Biodiesel Fuel from Vegetable Oil Using Immobilized *Candida Antarctica* Lipase. *Journal of the American Chemical Society,*7:4.

[24] Du W., Y. Xu, D. Liu, J. Zeng.(2004). Comparative study on lipase-catalyzed transformation of soybean oil for biodiesel production with different acyl acceptors. *Journal of Molecular Catalysis B: Enzymatic.*, 30: 3-4:125-129.

[25] Devanesan M.G., Viruthagiri T. and SugumarN..(2007). Transesterification of Jatropha oil using immobilized *Pseudomonas fluorescens. African Journal of Biotechnology.*, 6:21: 2497-2501.

[26] Iso M., Chen B., Eguchi M., Kudo T. and Shrestha S.(2001). Production of biodiesel fuel from triglycerides and alcohol using immobilized lipase. *Journal of Molecular Catalysis B: Enzymatic.,* 16:53–58.

[27] Bélafi-Bakó K., Kovács F., Gubicza L. and Hancsók_J. (2002). Enzymatic Biodiesel Production from Sunflower Oil by *Candida antarctica* lipase in a Solvent-free System. *Biocatalysis and Biotransformation.*20: 6:437-439.

[28] Lee D.H., Kim J.M., Kang S.W., Lee J.W., Kim S.W.(2006). Pretreatment of lipase with soybean oil before immobilization to prevent loss of activity. *Biotechnology Letters.*, 28:1965–1969.

[29] Chang H., Liao H., Lee C. and Shieh C. Optimized synthesis of lipase-catalyzed biodiesel by Novozym 435.(2005). *Journal of Chemical Technology and Biotechnology.*, 80:3:307–312.

[30] Xu Y., Du W., Zeng J. and Liu D. (2004). Conversion of Soybean Oil to Biodiesel Fuel Using Lipozyme TL IM in a Solvent-free Medium. *Biocatalysis and Biotransformation*, 22:1:45-48.

[31] Li L., Du W., Liu D., Wang L. and Li Z.(2006). Lipase-catalyzed transesterification of rapeseed oils for biodiesel production with a novel organic solvent as the reaction medium. *Journal of Molecular Catalysis B: Enzymatic.*,43:1-4:58-62.

[32] Awang R., Ghazuli M.R. and Basri M.(2007). Immobilization of lipase from *Candida rugosa* on palm-based polyurethane foam as a support material. *American Journal of Biochemistry and Biotechnology*, 3: 163-166.

[33] Huang Y., Zheng H. And Yan Y.(2010). Optimization of lipase-catalyzed transesterification of lard for biodiesel production using response surface methodology. *Applied Biochemistry and Biotechnology*, 160: 504-515.

[34] Hsu A., Jones K., Marmer W.N. and Fogolia T.A. (2001). Production of alkyl esters from tallow and grease using lipase immobilized in phyllosilicate sol-gel *Journal of the American Chemical Society.*78:6:585-588.

[35] Kumari A, Mahapatra P., GarlapatiV.K. and Banerjee R. (2009). Enzymatic transesterification of Jatropha oil. *Biotechnology for Biofuels.,* 2:1.

[36] Fukuda H., Hama S., Tamalampudi S. and Noda H.(2008). Whole-cell biocatalysts for biodiesel fuel production. *Trends in Biotechnology*, 26:12.

[37] Sun T., Du W. and Liu D. (2009). Prospective and impacts of whole cell mediated alcoholysis of renewable oils for biodiesel production. *Biofuels, Bioprod. Bioref.*, 3: 633–639.

[38] Nakashima T. (1990). Cell aggregation as a trigger for enhancement of intracellular lipase production by a *Rhizopus* species. *Journal of Fermentation and Bioengineering*.,70:83-89.

[39] Hama S., TamalampudiS.,FukumizuT., MiuraK., YamajiH., KondoA. and Fukuda H. (2006). Lipase localization in *Rhizopusoryzae* cells immobilized within biomass support particles for use as whole-cell biocatalysts in biodiesel-fuel production. *Journal of Biosciences and Bioengineering.* 101:4:328-333.

[40] Wang A., Mulchandani A. and Chen W. (2001). Whole-Cell Immobilization Using Cell Surface-Exposed Cellulose-Binding Domain. *Biotechnology Progress*, 17:3:407-411.

[41] Narita J., Okano K.,Kitao T., Ishida S., Sewaki T., Sung M., Fukuda H. and Kondo A. (2006). Display of α–amylase on the surface of Lactobacillus casci cclls by usc of thc PgsA anchor protcin, and production of lactic acid from starch. *Appl. Environ. Microbiol.*, 72:1:269–275.

[42] Matsumoto T., Fukuda H., Ueda M., Tanaka A. and Kondo A.(2002). Construction of Yeast Strains with High Cell Surface Lipase Activity by Using Novel Display Systems Based on the Flo1p Flocculation Functional Domain. *Applied and Environmental Microbiology*., 68:9:4517–4522.

[43] Shiraga S., Kawakami M., Ishiguro M. and Ueda M. (2005). Enhanced Reactivity of *Rhizopusoryzae*Lipase Displayed on Yeast Cell Surfaces in Organic Solvents: Potential as a Whole-Cell Biocatalyst in Organic Solvents. *Applied And Environmental Microbiology*,71: 8: 4335–4338.

[44] Jiang Z., Gao B., Ren R., Tao X., Ma Y. and Wei D. (2008). Efficient display of active lipase LipB52 with a *Pichiapastoris*cell surface display system and comparison with the LipB52 displayed on *Saccharomyces cerevisiae*cell surface. *BMC Biotechnology.*, 8:4.

[45] Adachi D., Hama S., Numata T., Nakashima K., Ogino C., Fukuda H. and Kondo A. (2011). Development of an *Aspergillusoryzae* whole-cell biocatalyst coexpressing triglyceride and partial glyceride lipases for biodiesel production. *BioresourceTechnology*. 102:6723–6729.

[46] Baek J.H., Han M., Lee S.H. and Lee S.Y.(2011). Enhanced Display of Lipase on the *Escherichia coli* Cell Surface based on Transcriptome Analysis. *Applied and Environmental Microbiology*. 76:3:971–973.

[47] Wu C.H., Mulchandani A. and Chen W. (2008). Versatile microbial surface-display for environmental remediation and biofuels production. *Trends in Microbiology*, 16:4:181-188.

[48] Hashemizadeh S.N., Tavakoli O., Tabandeh F., Karkhane A.A. and Forghanipour Z. (2011). A Comparative Study of Immobilized-Whole Cell and Commercial Lipase as a Biocatalyst for Biodiesel Production from Soybean Oil. *Bioenergy Technology. World Renewable Congress, Sweden.*

[49] Ban K., Kaieda M., Matsumoto T., Kondo A., Fukuda H.(2001). Whole cell biocatalyst for biodiesel fuel production utilizing *Rhizopusoryzae*cells immobilized within biomass support particles. *Biochemical Engineering Journal*,. 8:39-43.

[50] Chen J. and Li G. (2010). Optimization of Biodiesel Production Catalyzed by Fungus Cells Immobilized in Fibrous Supports. *Applied Biochemistry and Biotechnology*, 161:181–194.

[51] Li W., Du W. and Liu D. (2007). *Rhizopusoryzae* IFO 4697 whole cell catalyzed methanolysis of crude and acidified rapeseed oils for biodiesel production in *tert*-butanol system. *Process Biochemistry*, 42:11:1481-1485.

[52] Xiao M., Mathew S. and Obbard J.P.(2010). A newly isolated fungal strain used as whole-cell biocatalyst for biodiesel production from palm oil. *GCB Bioenergy*, 2: 45–51.

[53] Pazouki M., Zamani F., Zamzamian S.A.H., Najafpour G. (2011). Study on Reaction Conditions in Whole Cell Biocatalyst methanolysis of Pretreated Used Cooking Oil. Bioenergy Technology. *World Renewable Energy Congress- Sweden.*

[54] Ying M. and Chen G. (2007). Study on the production of biodiesel by Magnetic Cell Biocatalyst based on lipase producing *Bacillus subtilis*. *Applied Biochemistry and Bitechnology*, 7: 136-140.

[55] Demain AL. (1971). Overproduction of microbial metabolites and enzymes due to alteration of regulation. *Adv Biochem Eng*., 1:113–42.

[56] Betz JL, Brown PR, Smyth MJ, Clarke PH.(1974). Evolution in action. *Nature,* 247:261–4.

[57] An S. Y., Kim S.W., Choi Y.L., Cho Y.S., Joo W.H. and Lee Y.C. (2003). Cloning, expression in *Escherichia coli* and enzymatic properties of lipase from *Pseudomonas* sp. SW-3. *J. Microbiol.,* 41:95-101.

[58] Long J.D., Xu J.H., Zhao L.L., Pan J., Yang S. and Hua L.(2007). Over expression of *Serratiamarcescens*lipase in *E. coli.* For efficient bioresolution of recemicketoprofen. *J. Mol. Catal. B. Enzym.,* 27:105-110.

[59] Ramchuran S.O., Vargas V.A., Hatti-Kaul R. and Karlsson E.N.(2006). Production of a lipolytic enzyme originating from *Bacillus halodurans* LBB2 in the methylotrophic yeast *Pichiapastoris. Appl. Microbiol. Biotechnol,* 71:463-472.

[60] Yang J., Zhang B. and Yan Y.(2009). Cloning and expression of *Pseudomonas fluorescens* 26-2 lipase gene in *Pichiapastoris* and characterizing for transesterification. *Appl. Biochem. Biotechnol*, 159:2: 355-365.

[61] Yang J., Guo D. and Yan Y.(2006). Cloning, expression and characterization of a novel thermal stable and short-chain alcohol tolerant lipase from *Burkholderiacepacia*strain G63. *Journal of Molecular Catalysis B: Enzymatic,* 45:91-96.

[62] Palomo J.M., Munoz G., Fernandez Lorente G., Mateo C., Fernandez-Lafuente R. and Guisan J.M.(2002). Interfacial adsorption of lipase on very hydrophobic support (octadecyl-Sepabeds): immobilization, hyperactivation and stabilization of the open form of lipases. *J MolCatal B: Enzym.*,19–20:279–86.

[63] Foresti M.L, Alimentib G.A. and Ferreira M.L..(2005). Interfacial activation and bioimprinting of *Candida rugosa*lipase immobilized on polypropylene: effect on the enzymatic activity in solvent-free ethyl oleate synthesis. *Enzyme and Microbial Technology,* 36:338–349.

[64] Mingarro I., Abad C. and Braco L. (1995). Interfacial activation-based molecular bioimprinting of lipolytic enzymes. *Proc. Natl. Acad. Sci. USA (Biochemistry),*92(8): 3308-3312.

[65] Gonzalez-Navarro H. and Braco L. (1997). Improving lipase activity in solvent-free media by interfacial activation-based molecular bioimprinting. *Journal of Molecular Catalysis B: Enzymatic*. 3:111-119.

[66] Harris J.L. and Craik C.S.(1998). Engineering enzyme specificity. *CurrOpinChem Biol.* 2:127-132.

[67] Fishman A. and Cogan U. (2003). Bio-imprinting of lipases with fatty acids. *Journal of Molecular Catalysis B: Enzymatic*. 22:193–202.

[68] Bornscheuer U.T., Bessler C., Srinivas R. and Krishna S.H.(2002). Optimizing lipases and related enzymes for efficient application. *Trends in Biotechnology*. 20:10:433-437.

[69] Wahler D. and Reymond J.L. (2001). Novel methods for biocatalyst screening. *CurrOpinChem Biol,* 5:152-158.

[70] Cadwell R.C., Joyce G.F. (1992). Randomization of genes by PCR mutagenesis. *PCR Methods Appl.,* 2: 28–33.

[71] Stemmer W.P.C. (1994). Rapid evolution of a protein in vitro by DNA shuffling. *Nature*, 370:389–391.

[72] Scheib H., Pleiss J., Stadler P., Kovac A., Potthoff A. P., Haalck L., Spener F., Paltauf F. and Schmid R.D. (1998). Rational design of *Rhizopusoryzae*lipase with modifiedstereoselectivity toward triracylglycerols. *Protein Eng,* 11:675-682.

[73] Scheib H.,Pleiss J., Kovac A.,Paltauf F. and Schmid R.D.(1999). Stereoselectivity of *Mucorales*lipases toward triracylglycerols — a simple solution to a complex problem. *Protein Sci.*, 8:215-221.

[74] Liebeton K., Zonta A., Schimossek K., Nardini M., Lang D., Dijkstra B.W., Reetz M.T. and Jaeger K.E. (2000). Directed evolution of an enantioselectivelipase. *Chem. Biol.*, 7:709-718.

In: Advances in Biotechnology ISBN: 978-1-62417-842-9
Editors: H. K. Dhingra, P. N. Jha et al.

Chapter 5

Biological Control of Insect Pests Using Entomopathogenic Nematodes

***Jola Dubey**, *B. N. Tiwary and Sudershan Ganguly*[1]**

Department of Biotechnology, Guru Ghasidas Vishwavidyalaya, Bilaspur, Chhattisgarh, India

[1]Division of Nematology, Indian Agricultural Research Institute, New Delhi, India

Abstract

Entomopathogenic nematodes belonging to the families Steinernematidae and Heterorhabditidae have tremendous biocontrol potential against several insect pests. These nematodes carry mutualistic bacteria of the genus *Xenorhabdus* for Steinernematidae and *Photorhabdus* for Heterorhabditidae. The infective juveniles (IJs) or dauer juveniles containing bacterial symbiont infect the host through natural openings or inter- segmental areas and release the bacterium in hemocoel where it propagates, causes septicemia and kills the host within 24-48 h. It also protects the cadaver from colonization by other micro-organisms. These nematodes being compatible with recommended doses of several pesticides and non target microorganisms can suitably be incorporated in the integrated pest management (IPM) schedules. Naturally occurring entomopathogenic nematodes and their symbiotic bacteria are important biotic factor in suppression of insect pest

* Email: pandeyjola@rediffmail.com.

populations in soil and cryptic habitats. The virulent species of these nematodes are commercially produced as biological control agents all over the world encompassing North America, Europe, Asia and Australia in glasshouse crops, orchards, ornamentals, turf, lawn, and forestry. India has a great potential to exploit these beneficial nematodes for the suppression of insect pests. Recent emphasis on mass production and formulation technologies of these nematodes in India stresses a need to implement safer and effective pest control methods. This article provides an overview of recent development on formulation and commercialization of entomopathogenic nematodes, and evaluates their potential exploitation in India.

INTRODUCTION

The inappropriate and indiscriminate use of pesticides had led to serious environmental threat to human life (Anon., 2007, Walter *et al.*, 2005). Spray is being done even before one or two days of picking of green vegetables. These days most of raw food materials we consume are loaded with substantial quantities of pesticides which are used during the cultivation of the crops. According to WHO estimates approximately one million people are taken ill every year with pesticides poisoning and up to 20,000 of those die in agony (Miller, 2004). The use of chemical pesticides has gradually expanded, and the recent world insecticides sales (including acaricides), is estimated at 20-25 billion US $ per annum (Lisansky and Coombs, 1994). Besides their non target effects and hazardous nature, they are becoming more expensive and some are loosing their effectiveness because of the development of resistant strains (Mukhopadhyay, 1994). However, in recent era most of synthetic pesticides are non-biodegradable and their incessant applications pose problem to non-target organisms (Ambridge and Haines, 1987; Nichols *et al.*, 1993).

In recent years, a large number of synthetic pesticides including insecticides and fungicides have been banned in the western world because of their undesirable attributes such as high toxicity, long degradation periods, accumulation in the food chain and extension of their influence to destroy both useful and harmful pests (Barnard *et al.,* 1997). In the developing countries such as in India, these are still being used despite their detrimental effects (Wodageneh and Wulp, 1997). It has been estimated that hardly 0.1% of the agrochemicals used in crop protection reach the target pest, leaving the remaining 99.9% to enter the environment to cause hazards to non-target organisms including humans. On the other hand, many plant pathogenic microorganisms have developed resistance against known chemical pesticides (Williams and Heymann, 1998; Witte, 1998). In addition, investigation in the area of chemical control has been slowed down due to the rising cost of production and sales of the limited fungicides available in the market, which the poor subsistent farmers are unable to afford. Basic research for over more than 40 years in biology and biochemistry has made it possible to envisage not only how new pesticides may be synthesized but also a completely new approach for the protection of plants using secondary plant products or microorganisms which may be toxic to a specific pest yet harmless to man. India's vast potential of exporting agri-products will be successful only when we produce agri-products through minimum use of chemical pesticides or no pesticides.

Environmental pollution and toxic hazards due to haphazard use of chemical pesticides have forced the scientists to look for safe and environmental–friendly alternative strategies. Thus there is an all-round compulsion to go in for bio-rational alternative arsenals which can be eco-friendly and benign to environment. Bio-control agents are the best alternative available today. Microbial control envisages the use of bacteria, viruses, fungi and nematodes or their bi- products in disease and pest control. Research on the use of microorganisms i.e. Bacteria, Fungi, viruses and nematodes in disease and pest management during the past couple decades has led to the development of several microbial pesticides. It is now feasible to replace many of the highly toxic insecticides and fungicides by bio-pesticides which are safe and non hazardous to the environment, such as the use of microorganisms including entomopathogenic nematodes (EPN) (Bhaskaran *et al.*, 1994).

The term "entomopathogenic" is being widely used exclusively for steinernematid and heterorhabditid nematodes which along with their symbiotic bacteria kill the insect host quickly i.e. within a day or 2 after the invasion (Gaugler, 2006). EPNs have emerged as suitable biocontrol agents for eco-friendly pest management (Gaugler and Kaya, 1990; Kaya and Gaugler, 1993; Burnell and Stock, 2000; Gaugler, 2002).

These are soil-inhabiting, lethal insect parasitoids that belong to the Phylum Nematoda, commonly called roundworms. The term *entomopathogenic* comes from the Greek word "*entomon*", meaning insect, and "*pathogenic*", which means causing disease. Although many other parasitic nematodes cause diseases in plants, livestock, and humans, entomopathogenic nematodes, as their name implies, only infect insects and produce harmful effects on insect- pests. Entomopathogenic nematodes (EPNs) live inside the body of their host, and so they are designated as *endoparasitic*. These nematodes, though highly pathogenic, are facultative parasites due to their capability to complete their life-cycle on the symbiotic bacteria without the insect host.

However, the pathogenicity, host searching behaviour, and survivability of different nematode species are varied making them suitable in biological control programs. They infect many different types of soil insects, including the larval forms of butterflies, moths, beetles, and flies, as well as adult crickets and grasshoppers (Friedman, 1990). EPNs have been found in all inhabited continents and a range of ecologically diverse habitats, from cultivated fields to deserts. EPNs have tremendous biological control potential for managing a wide range of insect pests of crops as well as some household pests.

They have very high potentialities as bioagents against the insect pests, due to their several attributes – i) like parasitoids / predators, they have chemoreceptors and are motile, ii) like pathogens, they are highly virulent, killing their host within a day or two, iii) active host seeking, iv) can be cultured easily *in vitro*, v) have a high reproductive potential, vi) have broad host range, vii) safe to the plant and animal health, and the environment, viii) can be easily applied using standard spray equipments, ix) have potential to recycle in environment; compatible with many chemical pesticides, x) easy mass production, xi) long term efficacy.

LIFE-CYCLE OF EPN

Since they are economically important, the life-cycles of the genera Heterorhabditid and Steinernematid are particularly well known. Although not closely related, phylogenetically, both share similar life histories (Poinar, 1993). The cycle begins with an infective juvenile, whose only function is to seek out and infect new hosts. After entering an insect, infective juveniles release an associated mutualistic bacterium. These bacteria of the genus *Xenorhabdus* or *Photorhabdus*, for steinerernematides and heterorhabditids, respectively—cause host mortality within 48 hours. The nematodes provide shelter to the bacteria, which, in return, kill the insect host and provide nutrients to the nematode. Together, the nematodes and bacteria feed on the liquefying host, and reproduce for several generations inside the cadaver. Steinernematid infective juveniles may become males or females, whereas heterorhabditids develop into self-fertilizing hermaphrodites with later generations producing two sexes.

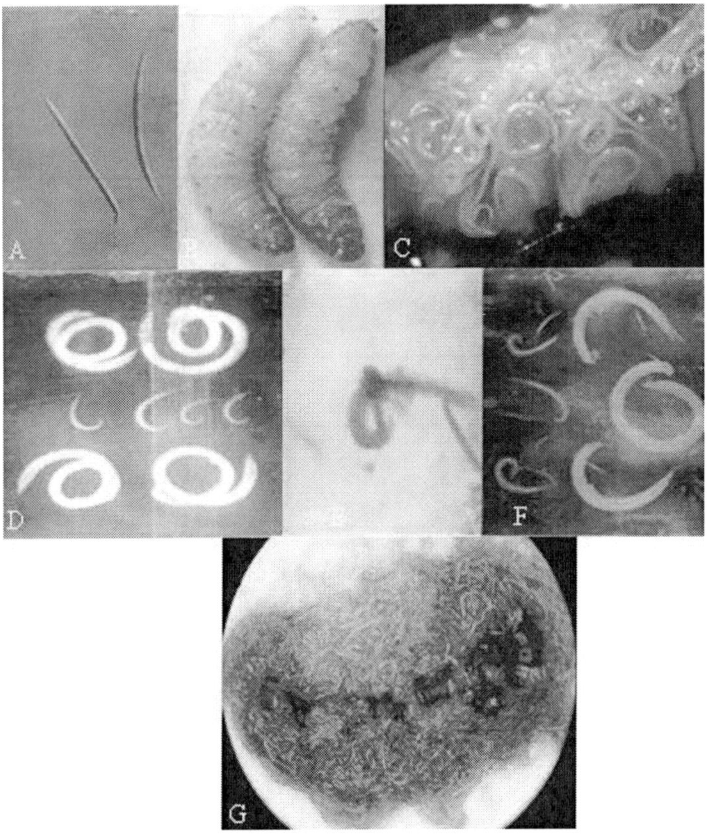

Figure 1. Life-Cycle of EPN (*Steinernema*). A: Infective Juveniles B: Entered and killed Insect cadaver C: Nematodes developing inside the insect body D: Ist generation adults E: Mating F: IInd generation adults G: Infective Juveniles emerge *en-masse* from insect cadaver body.

When food resources in the host become scarce, the adults produce new infective juveniles adapted to withstand the outside environment. After about a week, hundreds of thousands of infective juveniles emerge and leave in search of new hosts, carrying with them an inoculation of mutualistic bacteria, received from the internal host environment (Gaugler, 2006) (Figure 1).

Nematode-Bacterium-Insect Complex

These steinernematids and heterorhabditids have symbiotic relationships with the bacteria belonging to the genera *Xenorhabdus* and *Photorhabdus*, respectively, which is species specific, as for e.g. *X. indica* with *S. thermophilum; X. bovieni* with *S. feltiae*, *X. nematophilus* with *S. carpocapsae*, *X. poinarii* with *S. glaseri* and *Photorhabdus luminescens* with *Heterorhabditis bacteriophora.* Some of the symbiotic bacteria associated with different species of *Steinernema* and *Heterorhabditis* are given in the table 1.

Table 1. Symbiotic bacteria associated with different species of *Steinernema* and *Heterorhabditis.* (Modified after Boemare, 2002)

Nematode species	Symbiotic bacterium
Steinernema spp.	Xenorhabdus spp.
S. affine	*X. bovienii*
S. bicornutum	*X. budapestensis*
S. cubanum	*X. poinarii*
S. feltiae	*X. bovienii*
S. glaseri	*X. poinarii*
S. intermedium	*X. bovienii*
S. kraussei	*X. bovienii*
S. kushidai	*X. japonica*
S. longicaudatum	*X. beddingii*
S. rarum	*X. szentirmaii*
S. scapterisci	*X. innexi*
S. serratum	*X. ehlersii*
S. thermophilum	*X. indica*
S. carpocapsae	*X. nematophila*
Heterorhabditis spp.	Photorhabdus spp.
H. bacteriophora subgroup Brecon	*P. luminescens luminescens*
H. bacteriophora subgroup HP88	P. luminescens laumondii
H. indica	*P. luminescens akhurstii*
H. zealandica	*P. temperata*
H. bacteriophora subgroup NC	*P. temperata*
H. megidis Nearctic group (Ohio, Wisconsin)	*P. temperata*
H. megidis Palaearctic group	*P. temperata temperata*

These bacterial symbionts are gram-ve, anaerobic rods, classified under the family Enterobacteraceae. Excellent mutualism occurs between the nematode and the bacterium (Poinar, 1966; Kaya, 1993; Kaya and Gaugler, 1993).

The nematode is dependent on the bacterium for: Killing its host; creating a suitable environment for nematode development and preventing the infection by secondary microorganisms; Serving as a food source and Breaking down the host tissues to serve as nutrient source for nematode.

The bacterium needs the nematode for: Protection from the external environment, Penetration into host's haemocoel and Inhibition of the host's antibacterial proteins.

The nematode-killed insects are flaccid and do not have a putrid odor because the mutualistic bacteria produce antibiotics which prevent the growth of secondary microorganisms. These bacteria are known to exhibit peculiar behavior of phase transition (Akhurst, 1980; Akhurst and Boemare, 1990). Although, both phases are equally pathogenic to the insects, the phase one produces better conditions for nematode reproduction, an important consideration in commercial use of these microbes as bio- pesticides.

The infective juvenile (IJs) of EPN is a microscopic organism, 0.5 to 1.5 mm long, depending on species. The third stage juvenile of these nematodes have closed mouth and anus and cannot feed until it finds an insect. Usually it is found in soil and is activated by insect movement and then follows a gradient of CO_2 to find the insect larvae to get into the insect's blood cavity in order to kill it. EPNs enter through the insect's natural body openings, the mouth, anus or respiratory inlets (spiracles) and then penetrate into the blood cavity from the gut; *Heterorhabditis* species can also penetrate through chinks in the insect's armour (the inter skeletal membranes) by scratching away at these with a special tooth (Bedding and Molyneux, 1982). Once in the insect's blood, infective juvenile releases symbiotic bacterium (*Xenorhabdus* spp. in *Steinernema*, *Photorhabdus* spp. in *Heterorhabditis*) (Kaya and Gaugler, 1993).

The bacteria replicates logarithmically in the insect's rich haemolymph, then causes a suppression to the host immune capacity against both the symbiotic nematodes and the bacteria themselves by inhibiting phospholipase A of the insect host (Park and Kim, 2000; Park *et al.*, 2003) and convert the insect into suitable food. At this immuno-compromised state, the infected insect can be susceptible to other saprophytic pathogens. To maintain monoxenic condition, bacterium synthesizes and secretes broad spectral antibiotics as well as narrow spectral bacteriocins (Akhurst, 1982; Boemare *et al*., 1992) consequently inhibiting growth of other pathogens and preserves the dead insect from putrefaction while the nematodes feed and reproduce in it.

The monoxenic state leads to lethal septicemia of the target insect, which is required for development of the symbiotic nematodes in the cadaver. Various kinds of antibiotics against bacteria and fungi are synthesized and secreted during the bacterial growth of *Xenorhabdus* and *Photorhabdus* (Webster *et al*., 2002) Among them, indoles (Paul *et al*., 1981), dithiolopyrrolones (McInerney *et al*., 1991a) and xenocoumacins (McInerney *et al*., 1991b) are the antibacterial compounds synthesized by *Xenorhabdus* spp. The mechanism of antibacterial activity has not been elucidated, but primarily speculated as inhibition of RNA and protein synthesis by an accumulation of the regulatory nucleotide, guanosine-3 pyrophosphate (ppGpp) especially in indole compounds (Sundar and Chang, 1993). This antibiotic aspect of the entomopathogenic bacteria reflects the potential medical and agricultural importance of their metabolic products (Webster *et al*., 1998).

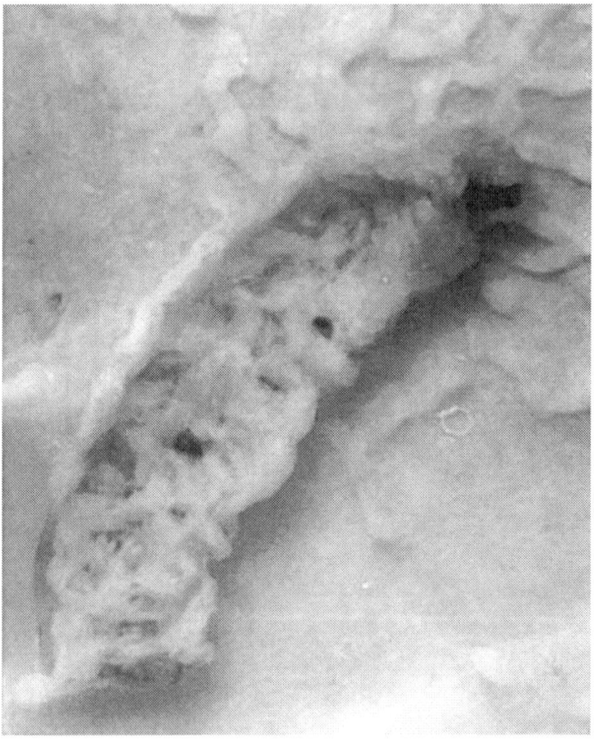

Figure 2. *En-masse* production of nematode progeny from the insect cadaver.

These bacteria kill the insect within a day or two and IJs emerge from the cadaver. From a medium-sized insect cadaver 100,000 to 500,000 IJs are produced which leave the decaying cadaver in 7-10 days and seek out new insect pest hosts (Figure 2).

FORAGING STRATEGIES

The foraging strategies of entomopathogenic nematodes vary with species, influencing their soil depth distributions and host preferences. Infective juveniles find their hosts either by ambushing or cruising foraging strategy (Campbell and Gaugler, 1997). In order to ambush prey, some *Steinernema* species nictate, or raise their bodies off the soil surface so they are better poised to attach to passing insects, which are much larger in size. Many *Steinernema* are able to jump by forming a loop with their bodies that creates stored energy which, when released, propels them through the air. Other species adopt a cruising strategy and rarely nictate.

Infective juveniles of entomopathogenic nematodes use three different strategies to find their insect hosts:

- ***Ambush foraging***
  Ambushers such as *Steinernema carpocapsae* and *S. scapterisci* are adapted to "sit and wait" strategy to attack highly mobile insects (billbugs, sod webworms, cutworms, mole-crickets and armyworms) when they come in contact at the surface of the soil. These nematodes do not respond to host released cues but infective juveniles of some *Steinernema* spp can stand on their tails (nictate) and easily infect passing insect hosts by jumping on them. Since highly mobile insects live in the upper soil or thatch layer, ambushers are generally effective in infecting more insects on the surface than deep in the soil.
- ***Cruise foraging***
  Cruiser nematodes such as *Heterorhabditis bacteriophora, H. megidis, Steinernema glaseri* and *S. kraussei* generally move actively in search of hosts and therefore, they are distributed throughout the soil profile and more effective against less mobile hosts such as white grubs and black vine weevils. Cruisers never nictate but respond to carbon dioxide released by insects as cues. They roam through the soil searching for potential hosts.

- ***Intermediate foraging***
  Some nematode species such as *Steinernema thermophilum, S. feltiae* and *S.riobrave* have adapted a strategy in between ambushing and cruising called an intermediate strategy to attack both the mobile and sedentary/less mobile insects at the surface or deep in the soil. *Steinernema feltiae* is highly effective against fungus gnats and mushroom flies whereas *S. riobrave* is effective against corn earworms, citrus root weevils and mole crickets.

These foraging strategies influence which hosts the nematodes infect. For example, ambush predators such as *Steinernema carpocapsae* infect more insects on the surface, while cruising predators like *Heterorhabditis bacteriophora* infect insects that live deep in the soil.

HOST RANGE

The entomopathogenic activity of steinernematid and heterorhabditid species has been documented against a broad range of insect pests in a variety of habitats (Kaya and Gaugler, 1993). They have been used inundatively in a number of high value cropping systems (Koppenhofer *et al.*, 2000). The nematode-bacterium complex kills insects so rapidly that the nematodes do not form the intimate, highly adapted, host-parasite relationship characteristic of other insect nematode associations. This rapid mortality permits the nematodes to exploit a range of hosts that spans nearly all insect orders, a spectrum of activity well beyond that of any other microbial control agent. In laboratory tests, *S. carpocapsae* alone infected more than 250 species of insects from over 75 families in 11 orders (Poinar, 1983). The nematodes attack a far wider spectrum of insects in the laboratory where host contact is assured, environmental conditions are optimal, and no ecological or behavioral barriers to infection exist. Some nematode species may search for hosts at or near the soil surface (e.g., *S. carpocapsae* and *S. scapterisci*), whereas others are adapted to search deeper in the soil profile (e.g., *H. bacteriophora* and *S. glaseri*). The former group has been referred to as “ambusher”, which remains nearly sedentary while waiting for the mobile surface dwelling hosts. The latter group has been referred to as “cruiser” which is highly mobile, responds strongly to long-range host chemical cues, and is therefore best adapted to find sedentary hosts (Grewal *et al.*, 1994). Because the symbiotic bacterium kills insects so quickly, there is no intimate host-parasite relationship as is characteristic for other insect-parasitic nematodes. Consequently, entomopathogenic nematodes are lethal to an extraordinarily broad range of insect pests in the laboratory. Field host range is considerably more restricted, with some species being quite narrow in host specificity.

POPULATION ECOLOGY

Competition and Coexistence

Inside their insect hosts, EPNs experience both intra and interspecific competition. Intraspecific competition takes place among nematodes of the same species when the number of infective juveniles penetrating a host exceeds the amount of resources available. Interspecific competition occurs when different species compete for resources. In both cases, the individual nematodes compete with each other indirectly by consuming the same resource, which reduces their fitness and may result in the local extinction of one species inside the host. Interference competition, in which species compete directly, can also occur. For example, a steinernematid species that infects a host first usually excludes a heterorhabditid species. The mechanism for this superiority may be antibiotics produced by *Xenorhabdus*, the symbiotic bacterium of the steinernematid. These antibiotics prevent the symbiotic bacterium of the heterorhabditid from multiplying. In order to avoid competition, some species infective juveniles are able to judge the quality of a host before penetration. The infective juveniles of *S. carpocapsae* are repelled by 24-hour-old infections, likely by the smell of their own species mutualistic bacteria (Grewal *et al.*, 1997).

Interspecific competition between nematode species can also occur in the soil environment outside of hosts. Millar and Barbercheck (2001) showed that the introduced nematode *Steinernema riobrave* survived and persisted in the environment for up to a year after its release. *S. riobrave* significantly depressed detection of the endemic nematode *H. bacteriophora*, but never completely displaced it, even after two years of continued introductions. *S. riobrave* had no effect on populations of the native nematode, *S. carpocapsae*, though, which suggests that coexistence is possible. Niche differentiation appears to limit competition between nematodes. Different foraging strategies allow two species to co-exist in the same habitat. Different foraging strategies separate the nematodes in space and enable them to infect different hosts.

EPNs also occur in patchy distributions, which may limit their interactions and further support coexistence.

Population Distribution

Entomopathogenic nematodes are typically found in patchy distributions, which vary in space and time, although the degree of patchiness varies between species (Lewis, 2002). Factors responsible for this aggregated distribution may include behavior, as well as the spatial and temporal variability of the nematodes natural enemies, like nematode trapping fungus. Nematodes also have limited dispersal ability. Many infective juveniles are produced from a single host could also produce aggregates. Patchy EPN distributions may also reflect the uneven distribution of host and nutrients in the soil. EPNs may persist as metapopulations, in which local population fragments are highly vulnerable to extinction, and fluctuate asynchronously. The metapopulation as a whole can persist as long as the rate of colonization is greater or equal to the rate of population extinction. The founding of new populations and movement between patches may depend on the movement of infective juveniles or the movement of infected hosts. Recent studies suggest that EPNs may also use non-host animals, such as isopods and earthworms for transport (Eng *et al.*, 2005).

Community Ecology

Parasites can significantly affect their hosts, as well as the structure of the communities to which they and their hosts belong (Minchella and Scott, 1991). Entomopathogenic nematodes have the potential to shape the populations of plants and host insects, as well as the species composition of the surrounding animal soil community. Entomopathogenic nematodes affect populations of their insect hosts by killing and consuming individuals. When more EPNs are added to a field environment, typically at concentrations of 250,000 individuals per square meter, the population of host insects measurably decreases. Agriculture exploits this finding, and the inundative release of EPNs can effectively control populations of soil insect pests in citrus, cranberries, turf grass, and tree fruit. If entomopathogenic nematodes suppress the population of insect root herbivores, they indirectly benefit plants by freeing them from grazing pressure. This is an example of a trophic cascade in which consumers at the top of the food (nematodes) exert an influence on the abundance of resources (plants) at the bottom. The idea that plants can benefit from the application of their herbivore's enemies is the principle behind biological control. Consequently, much of EPN biological research is driven by agricultural applications.

Examples of the top-down effects of entomopathogenic nematodes are not restricted to agricultural systems. Researchers at the Bodega Marine Laboratory examined the strong top-down effects that naturally occurring EPNs can have on their ecosystem (Strong *et al.*, 1996). In a coastal shrub land food chain the native EPN, *Heterorhabditis heplialus*, parasitized ghost moth caterpillars, and ghost moth caterpillars consumed the roots of bush lupine. The presence *H. heplialus* correlated with lower caterpillar numbers and healthier plants. In addition, the researchers observed high mortality of bush lupine in the absence of EPNs. Old aerial photographs over the past 40 years indicated that the stands where nematodes were prevalent had little or no mass die-off of lupine. In stands with low nematode prevalence, however, the photos showed repeated lupine die-offs. These results implied that the nematode, as a natural enemy of the ghost moth caterpillar, protected the plant from damage. The authors even suggested that the interaction was strong enough to affect the population dynamics of bush lupine (Strong *et al.*, 1996).

Not only do enomopathogenic nematodes affect their host insects, they can also change the species composition of the soil community. Many familiar animals like earthworms and insect grubs live in the soil, but smaller invertebrates such as mites, collembolans, and nematodes are also common. Aside from EPNs, the soil ecosystem includes predatory, bacteriovorous, fungivorous and plant parasitic nematode species. Since EPNs are applied in agricultural systems at a rate of 1,000,000 individuals per acre, the potential for unintended consequences on the soil ecosystem appears large. EPNs have not had an adverse effect on mite and collembolan populations, yet there is strong evidence that they affect the species diversity of other nematodes. In a golf course ecosystem, the application of *H. bacteriophora*, an introduced nematode, significantly reduced the abundance, species richness, maturity, and diversity of the nematode community. EPNs had no effect on free-living nematodes. However, there was a reduction in the number of genera and abundance of plant-parasitic nematodes, which often remain enclosed within growths on the plant root. The mechanism by which insect parasitic nematodes have an effect on plant parasitic nematodes remains unknown. Although this effect is considered beneficial for agricultural systems where plant parasitic nematodes cause crop damage, it raises the question of what other effects are possible. Future research on the impacts EPNs have on soil communities will lead to greater understanding of these interactions.

In aboveground communities, EPNs have few side effects on other animals. One study reported that *Steinernema felidae* and *Heterorhabditis megidis*, when applied in a range of agricultural and natural habitats, had little impact on non-pest arthropods. Some minimal impacts did occur, however, on non-pest species of beetles and flies. Unlike chemical pesticides, EPNs are considered safe for humans and other vertebrates.

RESPONSE TO DISTURBANCE

Frequent disturbance often perturbs agricultural habitats and the response to disturbance varies among EPN species. In traditional agricultural systems, tilling disturbs the soil ecosystem, affecting biotic and abiotic factors. For example, tilled soils have lower microbial, arthropod, and nematode species diversity. Tilled soil also has less moisture and higher temperatures. In a study examining the tolerances of different EPN species to tillage, the density of a native nematode, *H. bacteriophora*, was unaffected by tillage, while the density of an introduced nematode, *S. carpocapsae*, decreased. The density of a third nematode introduced to the system, *S. riobrave*, increased with tillage (Millar and Barbercheck, 2002). Habitat preferences in temperature and soil depth can partially explain the nematodes' different responses to disturbance. *S. carpocapsae* prefers to remain near the soil surface and so is more vulnerable to soil disturbance than *H. bacteriophora*, which forages deeper and can escape the effects of tillage. *S. riobrave* may have responded well to tillage because it is better at surviving and persisting in hotter and drier conditions created by tillage. The response of EPNs to other forms of disturbance is less well defined. Nematodes are not affected by certain pesticides and are able to survive flooding. The effects of natural disturbances such as fire have not been examined.

PRODUCTION, FORMULATION AND APPLICATION TECHNOLOGIES

It is far too expensive to rear EPNs by *in vitro* media as they require separate media source for nematodes growth and their associated bacterium with suitable fermentor. The *in vivo* process, however, lacks any economy of scale; the labour, equipment, and material (insect diet) costs increase as a linear function of production capacity. However, in India the first successful and commercial scale of mass production of EPN species; *H. indica*, *S. carpocapsae*, *S. thermophilum* and *S. glaseri* on larvae of greater wax moth, *Galleria mellonella* L. (Lepidoptera: Pyralidae) and formulation of these nematodes have been developed after three years of rigorous research by Multiplex Biotech Pvt. Ltd., Tumkur, Karnataka, India. This is a pioneered organisation commercializing these entomopathogenic nematodes reared *in vivo*, to increase their virulent and survival. *Galleria mellonella* are most commonly used host insect to mass produce these beneficial nematodes because of its rich nutrient source available in body and easy to multiply in economical semi-synthetic diet source containing wheat and corn flour based media. About 2.5 billion infective juveniles (IJs) of EPN are recommended to treat one hectare where the insect pests are highly infested on any crop (Shaprio-Ilan *et al.*, 2006). The latest formulation developed by Multiplex Biotech Pvt. Ltd., is comprised of hydro gel based semi solid cream with IJs suspension containing 1 million nematodes in a polythene pouch that can be readily mixed in a water spray tank without blocking the nozzles. Fifty such pouches are recommended to spray in one acre against the highly infested crop. The nematode shelf-life of 2 to 3 months was achieved at room temperature (27-28°C) with maximum of 80 % survival by increasing the nematodes metabolic rate by mixing silica powder (0.1 %). The company do have products of entomopathogenic nematodes that are currently marketed satisfactorily in the name of 'Soldier" recommended as soil application for soil dwelling insect pests and "Bouncer" as foliar application for leaf, flower, fruit, pods, stem and borer insect pests against agriculture, horticulture and forest insect pests. Entomopathogenic nematodes are often applied to sites and ecosystems that routinely receive other inputs that may interact with nematodes including chemical pesticides, surfactants (e.g., wetting agents), fertilizers, and soil amendments. Often it is desirable to tank mix one or more inputs to save time and money. Infective juveniles are tolerant of short exposures (2-8 h) to most agrochemicals including herbicides, fungicides, acaricides, and insecticides (Sankar, 2009; Prasad *et al.*, 2009). Some pesticides act synergistically with EPNs and therefore improve nematode efficacy in inundative applications. Nematodes are also compatible with most inorganic fertilizers when they applied inundatively.

Although the biological control industry has acknowledged EPNs since the 1980s, relatively little is understood about their biology in natural and managed ecosystems (Georgis, 2002). Nematode-host interactions are poorly understood, and more than half of the natural hosts for recognized *Steinernema* and *Heterorhabditis* species remain unknown (Akhurst and Smith, 2002). Information is lacking because isolates of naturally infected hosts are rare, so native nematodes are often baited using *Galleria mellonella*, a lepidopteran that is highly susceptible to parasitic infection. Laboratory studies showing wide host ranges for EPNs were often overestimates, because in a laboratory, contact with a host is assured and environmental conditions are ideal; there are no "ecological barriers" to infection (Kaya and Gaugler, 1993). Therefore, the broad host range initially predicted by assay results has not always translated into insecticidal success.

The lack of knowledge about nematode ecology has resulted in unanticipated failures to control pests in the field. For example, parasitic nematodes were found to be completely ineffective against blackflies and mosquitoes due to their inability to swim (Lewis *et al.*, 1998). Efforts to control foliage-feeding pests with EPNs were equally unsuccessful, because nematodes are highly sensitive to UV light and desiccation (Lewis *et al.*, 1998). Comparing the life-histories of nematodes and target pests can often explain such failures. Each nematode species has a unique array of characteristics, including different environmental tolerances, dispersal tendencies, and foraging behaviors (Lewis *et al.*, 1998). Increased knowledge about the factors that influence EPN populations and the impacts they have in their communities will likely increase their efficacy as biological control agents.

Entomopathogenic nematodes should be applied at the first sign that a pest population is initializing to cause damage. Reapplying nematodes depends on the success of the first nematodes released. Their survivorship and success are based on environmental condition, moisture and soil type and percentage of living nematodes actually released during the first application. Nematodes should be reapplied on 7 days intervals if damage continues. In order to ensure maximum effectiveness, it is crucial to apply them at the optimum environmental conditions needed for their better survival. Therefore, it is best to irrigate the target site, both before and after application because they need moist conditions to prevent desiccation and aid with movement to find hosts. Also, the best results are obtained when the relative humidity is high, ambient temperature is neither extremely hot or cold, soil temperature is between 10 to 35°C, soil is moist and direct sunlight is minimal. All of these factors help to prevent the nematodes from drying out and increase their survival and virulence.

PESTICIDES SUSCEPTIBILITY

Growers will not adopt biological agents that do not provide efficacy comparable with standard chemical insecticides. Technological advances in nematode production, formulation, quality control, application timing and delivery, and particularly in selecting optimal target habitats and target pests, have narrowed the efficacy gap between chemical and nematode agents. Nematodes have consequently demonstrated efficacy in a number of agricultural and horticultural market segments.

Entomopathogenic nematodes are remarkably versatile in being useful against many soil and cryptic insect pests in diverse cropping systems, yet are clearly underutilized. Like other biological control agents, nematodes are constrained by being living organisms that require specific conditions to be effective. Thus, desiccation or ultraviolet light rapidly inactivates insecticidal nematodes; chemical insecticides are less constrained. Similarly, nematodes are effective within a narrower temperature range than chemicals, and are more impacted by suboptimal soil type, thatch depth, and irrigation frequency (Georgis and Gaugler, 1991). Nematode-based insecticides may be inactivated if stored in hot vehicles, cannot be left in spray tanks for long periods, and are incompatible with several agricultural chemicals. Certain species cannot be applied with high-pressure application equipment; unused nematodes cannot be applied the following year; different species require different screen sizes. Chemicals also have problems (e.g., mammalian toxicity, resistance, groundwater pollution, etc.) but a large knowledge base has been developed to support their use. Accelerated implementation of nematodes into IPM systems will require users to be more knowledgeable about how to use them effectively.

Infective juveniles are compatible with most but not all agricultural chemicals under field conditions. *Heterorhabditis bacteriophora* is the most sensitive among entomopathogenic nematodes to physical stress, yet even this species was tolerant to three days of laboratory immersion in 58 of 75 fungicides, herbicides, insecticides, and nematicides assayed (Rovesti *et al.*, 1988). Moreover, many of the chemicals noted to be toxic had a transient effect only, as the nematodes recovered quickly when the exposure ended. Thus, tank mixes and simultaneous application of nematodes with most pesticides and fertilizers should be practical when exposures are short. Chemicals to be used with care or avoided include the insecticides bendiocarb, chlorpyrifos, ethoprop, and isazophos, the fungicides anilazine, dimethyl benzyl ammonium chloride, fenarimol, and mercurous chloride, the herbicides 2, 4-D and trichlopyr, and the nematicide fenamiphos.

Commercial Availability

It has also been established that EPNs are safe from the point of view of human health and environment and hence exempted from registration in USA and some other countries (Ehlers, 2005). In some of the developed nations, EPNs are already being used commercially for managing insect pests of turf grass, fruit trees, mushrooms and other field crops. However, India is yet to exploit this area of high biological control potential.

The first nematode (*S. carpocapsae*) used successfully in the control of an insect pest was reported 30 years ago from Australia; Commonwealth Scientific and Industrial Research Organization (CSIRO) was the first in the world to use EPNs commercially against black vine weevil in ornaments and against currant borer moth in black currants (Georgis, 2002). There are several species of EPNs used around the world against a variety of pests in niche markets (e.g. fungus gnats in nurseries, hydroponics, and mushroom; weevils on ornamentals, strawberries, cranberries, citrus and bananas; scarabs of turf, ornamentals and blue berries; cutworms, webworms, billbugs and mole crickets; termites in wooden articles and trees; peach borer moth in apples and carpenter worm in shade trees in China, and fig trees in USA) (Poinar, 1990). Recently, out of 13% bio-insecticide sale in industrialized countries, EPN sale was only second to *Bacillus thuringiensis* at 80% (Lisansky and Coombs, 1994).

Considerable progress has also been made during the last 20 years on the subject dealing with taxonomy, biology, genetics, ecology, host range, production, application technologies, laboratory and field trials and commercialisation of EPNs and their symbiotic bacteria resulting in over 2000 publications. The entomopathogenic activity of steinernematid and heterorhabditid species has been documented against a broad range of insect pests in a variety of habitats. They have been used inundatively in a number of high value cropping systems. Various tests against mammals (mice, rabbits and monkeys) have shown that the EPNs tested were harmless when fed, injected or inhaled. They are also harmless to earthworms and other non-target organisms including plants and they are of course nonpolluting. Nematodes have been applied successfully against soil inhabiting insects (as soil application) as well as above-ground insects (foliar spray) in cryptic habitats (Arthers *et al.*, 2004; Shapiro-Ilan *et al.* 2006).

These nematodes have been produced in large scale in various countries for over 10 years and large numbers of production workers have been exposed without any adverse effects being recorded. The Environment Protection Agency (EPA) in the India, USA, Australia and many European countries has exempted EPNs from registration. When environmental benefits including safety for humans and other non-target organisms, reduction of pesticide residues in food, increased activity of other natural enemies, and increased biodiversity in managed ecosystems are taken into account, the advantages are numerous.

In India, work on EPNs started in 1966 when Rao and Manjunath, suggested the use of DD-136 strain of *Steinernema carpocapsae* for managing the insect pests of rice, sugarcane and apple. Thereafter, several researchers tested the bioefficacy of some exotic strains of *S. carpocapsae*, *S. glaseri*, *S. feltiae* and *H. bacteriophora* imported by them. In many cases, these nematodes yielded inconsistent results in field trials, probably due to their poor adaptability to the local agro-climatic conditions (Rahman *et al.*, 2000).

Although, bioefficacy research has been conducted in India with selected nematodes since mid-1960s, research programs dealing with other aspects of EPNs started only in 1998 at the Indian Agricultural Research Institute in New Delhi. Besides the New Delhi facility, other laboratories that focus on entomopathogenic nematodes include Project Directorate of Biological Control (PDBC, Bangalore), Gujarat Agricultural University (GAU), Anand and Tamil Nadu Agricultural University (TNAU), Coimbatore.

Till date, there are two genera listed under Steinernematidae – *Steinernema* and *Neosteinernema*, with 67 species in the former and only one species in the latter (Divya and Sankar, 2009). Under Heterorhabditidae, there is only one genus – *Heterorhabditis* with only 18 known species. Of these, only two species have been contributed from India: *Heterorhabditis indica* by Poinar *et al.* (1992) and *Steinernema thermophilum* by Ganguly and Singh (2000). The *S. thermophilum* has high biocontrol potential against several insect pests and prefers high temperature conditions (30-35°C), therefore, may prove to be useful in integrated pest management in tropical and subtropical parts of the world. Some of the important EPN species belonging to *Steinernema* and *Heterorhabditis* with their original localities and sources of isolation are listed in the Table 2.

Indigenous EPN species of *Steinernema* and *Heterorhabditis* isolated from Indian natural source with their original localities are listed in the Table 3.

Table 2. EPN species of *Steinernema* and *Heterorhabditis* with their original localities and sources of isolation (Divya and Sankar, 2009)

NEMATODE SPECIES		Original locality	Original source
Heterorhabditis Poinar 1976			
1.	*H. argentinensis* Stock 1993	Rafaela, Argentina	*Graphognathus* sp.
2.	*H. bacteriophora* Poinar 1976	Brecon, South	*Heliothis punctigera*
3.	*H. brevicaudis* Liu 1994	Australia	Soil
4.	*H. hawaiiensis* Gardener *et al.* 1994	Fujian Province,	Soil
5.	*H. indica* Poinar *et al.* 1992	China	Soil
6.	*H. marelata* Liu and Berry 1996	Hawaii, USA	Soil
7.	*H. megidis* Poinar *et al.* 1987	Coimbatore, India	Popillia japonica
8.	*H. zealandica* Poinar 1990	Seaside, Oregon, USA	Heteronychus arator
		Jeromesville, USA	
		Auckland,	
		NewZealand	
Steinernema Travassos 1927			
1.	*S. abbasi* Elawad *et al.* 1997	Sultanate of Oman	Soil
2.	*S. affine* (Bovien 1937) Wouts *et*	Denmark	Bibio sp.
	*al.*1982	Central Russia	Soil
3.	*S. arenarium* Wouts *et al.* 1982	Strazilovo, Yugoslavia	Soil
4.	*S. bicornutum* Tallosi *et al.* 1995	Czechoslovakia	Cydia pomonella
5.	*S. carpocapsae* (Weiser 1955)Wouts *et*	China	Soil
	al. 1982	Jining Province, China	Soil
6.	*S. caudatum* Xu *et al.* 1991	Western Cuba	Soil
7.	*S. ceratophorum* Jian *et al.* 1997	Denmark	Agrotis feltiae
8.	*S. cubanum* Mracek *et al.* 1994	New Jersey, USA	Popillia japonica
9.	*S. feltiae* (Filipjev 1934) Wouts *et*	South Carolina, USA	Soil
	*al.*1982	Central Province,	Soil
10.	*S. glaseri* (Steiner 1929) Wouts *et al.*	Kenya	Cephaleia abietis
	1982	Germany	Anomala cuprea
11.	*S. intermedium* (Poinar 1985) Mamiya	Hamikita,	Soil
	1988	Guangdong, China	Soil
12.	*S. karii* Waturu *et al.* 1997	Republic of Korea	Neocurtilla hexadactylla
13.	*S. krausei* (Steiner 1923) Travassos	LaCrosse, Florida,	Soil
	1927	USA	Soil
14.	*S. kushidai* Mamiya 1988	Oregon, USA	Soil
15.	*S. longicaudatum* Shen and Wang 1991	Puerto Rico	Soil
16.	*S. monticolum* Stock *et al.* 1997	Cordoba, Argentina	Soil
17.	*S. neocurtillae* Nguyen and Smart 1992	Waslaco, Texas, USA	Scapteriscus vicinus
18.	*S. orgonense* Liu and Berry 1996	Cordoba, Argentina	Soil
19.	*S. puertoricense* Roman and	Rivera, Uruguay	
	Figueroa1994	IARI Farm, Delhi,	
20.	*S. rarum* (de Doucet 1986) Mamiya	India	
	1988		
21.	*S. riobrave* Cabanillas *et al.* 1994		
22.	*S. ritteri* de Doucet and Doucet 1990		
23.	*S. scapterisci* Nguyen and Smart 1990		
24.	*S. thermophilum* Ganguly and Singh,		
	2000		

Table 3. *Steinernema* and *Heterorhabditis* species isolated from original localities and their sources in India (Divya and Sankar, 2009)

Nematode species	Original locality	Natural source (soil and insect)	References
Heterorhabditis sp.	Dulhera (UP)	Mango and sugarcane	Rishi Pal *et al.,* 2008
Heterorhabditis indica	Pallavpuram (UP)	Guava	Rishi Pal *et al.,* 2008
Steinernema sp.	Horticulture Nursery, Modipuram (UP)	Guava, mango	Rishi Pal *et al.*, 2008
Heterorhabditis sp.	University Old campus, Modipuram (UP)	Stevia, lemon, okra, mango	Rishi Pal *et al.*, 2008
Steinernema sp.	University Old campus, Modipuram (UP)	lemon, okra, mango	Rishi Pal *et al.,* 2008
Steinernema masoodi	IIPR, Kanpur	Soil	Ali *et al.,* 2005
Steinernema seemae	IIPR, Kanpur	Soil	Ali *et al.*, 2005
Steinernema siamkayai	IARI, New Delhi	Soil	Ganguly *et al.*, 2005
Steinernema riobrave	Gujrat	Soil	Ganguly *et al.*, 2002
S. thermophilum	IARI, New Delhi	soil	Gangly and Singh, 2000
Steinernema sp.	PDBC, Bangalore	soil	Hussaini *et al.*, 2001
S. bicornutum S. carpocapsae *H. indica*	Karnataka	-	Hussaini *et al.*, 2001
Steinernema spp. (10 populations) *Heterorhabditis* spp (7populations)	Assam, Delhi, Gujarat, H.P., U.P.	-	Kaushal *et al.*, 2000
Steinernema spp (11 populations) *Heterorhabditis* spp. (4 populations)	Gujarat	-	Vyas *et al.*,1998
Heterorhabditis sp.	Melur and Chekkanurnai, T.N.	-	Bhaskaran *et. al.*, 1994
Heterorhabditis sp.	Melur and Chekkanurnai, T.N.	-	Bhaskaran *et. al.*, 1994
H. indica	SBRI, Coimbatore (TN)	Sugarcane stem borer	Poinar *et al.* 1992
S. feltiae	ICRISAT, A.P.	-	Singh et al., 1992
H. bacteriophora	TNAU,Coimbatore	Soil	Sivakumar *et al.*, 1989
Steinernema sp.	-	-	Singh, 1977
Steinernema sp.	-	-	Mathur *et al.*, 1966

Most of the exotic strains of EPN are cold-adapted and have poor adaptability in the local soils, as already experienced by the Indian researchers. Therefore, for exploiting the biocontrol potential of EPN for insect pest management, there is an urgent need to isolate, characterize and document some more indigenous strains best adopted to the local agro-climatic conditions.

India being a mega-biodiversity sub-continent due to its diverse agro-climatic and agro-ecological conditions shows a great diversity of EPNs existing in this region. The strains isolated from different regions of this country will not only cater to the needs of Indian farmers, but also several other tropical and sub-tropical parts of the world.

Of the nearly 78 steinernematid and heterorhabditid nematodes identified to date, seven species are commercially available. Most of the suppliers are US based (Dunn and Smart, 1997), (Hunter, 1994). Comparison-shopping is recommended as prices vary greatly among suppliers. Caution is further advised with regard to application rates. One billion nematodes per acre (250,000 per m^2) is the rule-of-thumb against most soil insects (containerized and greenhouse soils tend to be treated at higher rates). Broadcast application rates suggested by some suppliers of 100 million per acre or less are not credible and not supported by controlled studies. A final caveat is that, just as one must select the appropriate insecticide to control a target insect, so must one choose the appropriate nematode species or strain. Suppliers should be asked about field tests supporting their recommended matching of insect target and nematode.

CHARACTERISTICS OF KEY SPECIES

- *Steinernema thermophilum: Steinernema thermophilum* Ganguly and Singh, 2000 is the first new species of this genus described from India. Over the last ten years, a lot of research has been carried out at Indian Agricultural Research Institute, New Delhi to make this nematode a commercially viable bio-control agent. This species has a wide host range since it can infect several insect species belonging to 6 orders. It acts at a wide range of temperature (20-35°C) and soil moisture (3-19% w/w) and can access the host located at soil surface as well as up to 10 cm soil depth, thus indicating its intermediate foraging behaviour (Gawas *et al.*, 2004). It was also found to be performing equally well on smooth and rough surfaces. Being indigenous, it is well adapted in local agro climatic conditions. The symbiotic bacterium associated with *S. thermophilum*, has been described as a new species of genus *Xenorhabdus,* i.e. *Xenorhabdus indica* sp.nov. The nearest phylogenetic relatives of this bacterium as per 16s rDNA similarity are *Xenorhabdus poinarii* and *Xenorhabdus szentirmaii* (97.7% each). Strains of the species were different from all the members of the genus in their ability to cause weak starch hydrolysis, being negative for Tween 60 and for adipic acid utilization, and as the only member of the genus to utilize but do not acidify esculin, acidify but not utilize ribose and 5-ketogluconate. Field efficacy of *Steinernema thermophilum* as foliar spray against diamond back moth (DBM) (*Plutella xylostella*) infesting cabbage was also tested. Cabbage plants with artificially inoculated DBM larvae were sprayed with *S. thermophilum* IJs at one of three concentrations (1000, 2000 or 3000 IJs/ml). The average of two years showed that *S. thermophilum* at 3000 IJs/ml caused the highest mortality of 46%, whereas at 2000 IJs/ml and the insecticide treatment caused 40.5 and 40% mortality in DBM larvae, respectively. The infectivity of IJs on DBM larvae increased with increasing IJ concentrations. The IJs treatments reduced the crop damage by 9.7 to 43.1%, whereas in the insecticide treatment, the reduction was 49.5%. The success story of *S. thermophilum* has now become a reality as it has been licensed to M/s Multiplex Biotech Pvt. Ltd. for its availability to the end users. It will also cater to the needs of other tropical and sub-tropical nations for incorporating in the integrated pest management schedules against several insect pests.

- *Steinernema carpocapsae:* The most studied, available, and versatile of all entomopathogenic nematodes. Important attributes include ease of mass production and ability to formulate in a partially desiccated state that provides several months of room-temperature shelf-life. Particularly effective against lepidopterous larvae, including various webworms, cutworms, armyworms, girdlers, and wood-borers. This species is a classic sit-and-wait or "ambush" forager, standing on its tail in an upright position near the soil surface and attaching to passing hosts. Consequently, *S. carpocapsae* tends to be most effective when applied against highly mobile surface-adapted insects. Highly responsive to carbon dioxide once a host has been contacted, the spiracles are a key portal of host entry. It is most effective at temperatures ranging from 22 to 28°C.

- *Steinernema feltiae:* Attacks primarily immature dipterous insects, including mushroom flies, fungus gnats, and tipulids. This nematode is unique in maintaining infectivity at soil temperatures below 10°C. *S. feltiae* offers lower stability than other steinernematids.
- *Steinernema glaseri:* The largest entomopathogenic nematode at twice the length but eight times the volume of *S. carpocapsae* infective juveniles. Attacks coleopterous larvae, particularly scarabs. This species is a cruise forager, neither nictating nor attaching well to passing hosts, but highly mobile and responsive to long-range host volatiles. Thus, this nematode is best adapted to parasitize hosts possessing low mobility and residing within the soil profile. Field trials, particularly in Japan, have shown that *S. glaseri* can provide control of several scarab species. Large size, however, reduces yield, making this species significantly more expensive to produce than other species. A tendency to occasionally "lose" its bacterial symbiote is bothersome. Moreover, the highly active and robust infective juveniles are difficult to contain within formulations that rely on partial nematode dehydration. In short, additional technological advances are needed before this nematode is likely to see substantial use.
- *Steinernema kushidai:* Only isolated so far from Japan and only known to parasitize scarab larvae. Although still in development as of this writing, early field trials against scarab larvae in Japan and the U.S. have been highly promising. Good laboratory stability and field persistence have been demonstrated. This small nematode is likely be see rapid implementation if quirky problems encountered in mass rearing can be resolved.
- *Steinernema riobravis:* This novel and highly pathogenic species, isolated to date only from the Rio Grande Valley of Texas, possesses several novel features. Its effective host range runs across multiple insect orders. This versatility is likely due in part to its ability to exploit aspects of both ambusher and cruiser means of finding hosts. Trials have demonstrated its effectiveness against corn earworm and mole crickets. In Florida, 60,000 acres of citrus are treated annually for control of citrus root weevil with impressive results. This is a high temperature nematode, effective at killing insects at soil temperatures above 35°C. Persistence is excellent even under semi-arid conditions, a feature no doubt enhanced by the uniquely high lipid levels found in infective juveniles. Its small size provides high yields whether using

in vivo (up to 375,000 infective juveniles per wax moth larvae) or *in vitro* methods. Only formulation improvements that impart increased stability are needed for this parasite to achieve its full potential.
It must also be noted that *S. riobravis* is currently being marketed for suppression of plant parasitic nematodes infesting turfgrass. There is substantial correlative data suggesting that some entomopathogenic nematodes can suppress plant species. Some skepticism may be healthy until this puzzling assertion can be fully confirmed by rigorously designed, multiple field experiments.

- *Steinernema scapterisci:* The only entomopathogenic nematode to be used in a classical biological control program, *S. scapterisci* was isolated from Uruguay and first released in Florida in 1985 to suppress an introduced pest, mole crickets. The nematode become established and presently contributes to control. *Steinernema scapterisci* is highly specific to adult mole crickets. Its ambusher approach to finding insects is ideally suited to the turfgrass tunneling habits of its host. Commercially available since 1993, this nematode is also sold as a biological insecticide, where its excellent ability to persist and provide long-term control contributes to overall efficacy. Availability is severely restricted due to the small market niche this nematode occupies. This is aggravated by the specificity of this nematode for a host that is very difficult to rear, precluding *in vivo* production.
- *Heterorhabditis bacteriophora:* Among the most important entomopathogenic nematodes, *H. bacteriophora* possesses considerable versatility, attacking lepidopterous and coleopterous insect larvae among other insects. This cruiser species appears most useful against root weevils, particularly black vine weevil where it has provided consistently excellent results in containerized soil. A warm temperature nematode, *H. bacteriophora* shows reduced efficacy when soil drops below 20°C. Characteristic poor stability has limited the usefulness of this interesting nematode: shelf-life is problematic and most infective juveniles persist only a few days following field release.
- *Heterorhabditis megidis:* First isolated in Ohio, this nematode is marketed in Western Europe for control of black vine weevil and various other soil insects. Its large size, characteristic heterorhabditid instability, and dearth of field efficacy data limit its utility at present.

CONCLUSION

Entomopathogenic nematodes belonging to the families Steinernematidae and Heterorhabditidae have tremendous biocontrol potential against several insect pests of crops. In USA, UK, Australia and some other developed countries, use of EPN has already been commercialized and there are more than 80 companies which are mass producing, formulating and marketing these nematodes, for managing several insect pests (root weevils, sciarid flies, grubs, mole crickets, etc.) of strawberries, cranberries, citrus, mushrooms and turf grass, etc. These nematodes being compatible with recommended doses of several pesticides and non target microorganisms can suitably be incorporated in the integrated pest management schedules. Although chemical insecticides have been the primary means of controlling soil insects, for many years, concern about public safety, environmental contamination, and reduced efficacy due to possible microbial degradation or insect tolerance and resistance have created a need for alternative control strategies. The great advantages of using entomopathogenic nematodes as biopesticides for insect pest control are in operator and end-user safety, absence of withholding periods, the advantage of minimizing the treated area by monitoring insect populations, minimal harm to natural enemies and lack of environmental pollution. The establishment of entomopathogenic nematodes would bring several benefits such as firmly establishing insect nematodes will promote the sustained use of agriculture and develop a better understanding of biodiversity. The potential of entomopathogenic nematodes (steinernematids and heterorhabditids) for insect control has been now fully realized and well documented, and the number of researchers studying them has also increased considerably during the last one decade. Till to date, there are 67 valid species of *Steinernema*, one species of *Neosteinernema* and 18 species of *Heterorhabditis*. Besides, there are several strains isolated from different parts of the world, which are yet to be identified. The accurate identification of the strains must be done before taking up any further research on them. Therefore, there is an urgent need to have strong biosystematics base in every country for the identification of entomopathogenic nematodes. Morphological features of infective juveniles and males offer the best differentiating characters but the identification of these nematodes by these standard morphological criteria is rarely straightforward and for this reason scientists turned to gel diffusion, immuno-electrophoresis, enzyme analysis and DNA (genomic and mitochondrial) based molecular biological techniques. The isozyme analysis technique is simpler, less time-consuming and less costly and therefore can be widely applied for identifying the samples obtained from large surveys. Polymerase chain reaction made it possible to identify nematodes using small sample size and now Internal transcribed spacer region (ITS) of ribosomal DNA repeat unit is preferably used for molecular taxonomy as well as identification and characterization. Numerous genetic and systematic studies have also focused on mtDNA. Conclusively, integration of morphological data together with DNA analysis is the best approach for the biosystematics of entomopathogenic nematodes and for accurate identification of the species, and this will have far reaching impact on all round developments in research for exploiting these bioagents in integrated pest management.

Commercialization of EPN, in future, will also promote the cottage industry for mass production of EPN, which will improve the socio-economic status of the farming community in general and farm women in particular. The advanced research on EPN will promote their use in IPM schedules for managing the insect pests of crops. It will minimize the use of chemical pesticides, thereby reducing the environmental and health hazards. Widespread use of EPN will also safeguard the soil health, which otherwise is losing its fertility due to chemicals.

Acknowledgement

The authors would like to thank Dr. R.K.S. Tiwari, Principal Scientist, Department of Plant Pathology, IGKV, Bilaspur, Chhattisgarh, for critical reading of the manuscript and for his valuable scientific and linguistic suggestions and editing.

References

Akhurst, R.J. (1980). Morphological and functional dimorphism in *Xenorhabdus* spp., bacteria symbiotically associated with the insect pathogenic nematodes *Neoaplectana* and *Heterorhabditis*. *J. Gen. Microbiol.* 121: 303-309.

Akhurst, R.J. (1982). Antibiotic activity of *Xenorhabdus* spp., bacteria symbiotically associated with insect pathogenic nematodes of the families *Heterorhabditidae* and *Steinernematidae*. *J. Gen. Microbiol.* 128: 3061-3065.

Akhurst, R.J. and Boemare, M.E. (1990). Biology and taxonomy of *Xenorhabdus*. In: *Entomopathogenic nematodes in biological control* (Eds. R. Gaugler and H.K. Kaya), CRC Press, Boca Raton, Florida, USA, pp 75-92.

Akhurst, R. and Smith, K. (2002). Regulation and safety. In: *Entomopathogenic Nematology* (Ed. R. Gaugler), CABI Publishing, New Jersey, pp 311-332.

Ambridge, E.M. and Haines, T.H. (1987). Some aspects of pesticide use and human safety in South East Asia. In: *Proceedings of 11th International Congress of Plant Protection*, Magallona, E.D., Manila, Philippines, pp 219-224.

Anonymous (2007). What is a pesticide? US Environmental (July 24, 2007). epa.gov. Retrieved on September 15, 2007, pp 98.

Arthurs, S., Heinz, K.M. and Prasifka, J.R. (2004). An analysis of using entomopathogenic nematodes against above-ground pests. *Bull. Entomol. Res.* 94: 297–306.

Barnard, C., Padgitt, M. and Uri, N.D. (1997). Pesticide use and its measurement. *International pest control.* 39: 161-164.

Bhaskaran, R.K.M., Sivakumar, C.V. and Veenugopal, M.S. (1994). Biocontrol potential of entomopathogenic nematode in controlling red hairy caterpillar *Amsacta albistriga* (Lepiddoptera: Archtiidae) of groundnut. *Indian J. Agric. Sci.* 64: 655-657.

Boemare, N.E., Boyer-Giglio, M.H., Thaler, J.O., Akhurst, R.J. and Brehelin, M. (1992). Lysogeny and bacteriocinogeny in *Xenorhabdus nematophilus* and other *Xenorhabdus* spp. *Appl. Environ. Microbiol.* 58: 3032-3037.

Burnell, A.M. and Stock, S.P. (2000). *Heterorhabdits*, *Steinernema* and their bacterial symbionts – lethal pathogens of insect. *Nematology* 2: 31-42.

Campbell, J.F. and Gaugler, R.R. (1997). Inter-specific variation in entomopathogenic nematode foraging strategy: dichotomy or variation along a continuum? *Fundamental and Applied Nematology* 20: 393-398.

Divya, K. and Sankar, M. (2009). Entomopathogenic nematodes in pest management. *Indian J. Sci.Tech.* 2: 53-60.

Dunn, R.A. and Smart, G.C. (1997). Suppliers of beneficial nematodes. Nematology Pointer No. 45 (SS-ENY-27), Fl. Coop. Ext. Serv., IFAS, Univ. Florida. pp 67-78.

Ehlers, R.U. (2005). Forum on safety and regulation. In: *Nematodes as biocontrol agents* (Eds. P.S. Grewal, R.U. Ehlers and D.I. Shapiro-Ilan) CABI Publishing, Wallingford, UK, pp 107-114.

Friedman, M.J. (1990). Commercial production and development. In: *Entomopathogenic nematodes in biological control* (Eds. R. Gaugler and H.K. Kaya), CRC Press, Boca Raton, Florida, USA, pp 153–172.

Ganguly, S. and Singh, L.K. (2000). *Steinernema thermophilum* sp. n. (Rhabditida: Steinernematidae) from India. *Int. J. Nematol.* 10: 183-191.

Gaugler, R. and Kaya, H.K. (1990). *Entomopathogenic nematodes in biological control.* CRC Press, Boca Raton, Florida, pp 365.

Gaugler, R. (2002). *Entomopathogenic Nematology.* Wallingford, UK: CABI publishing, pp 388.

Gaugler, R. (2006). Nematodes-Biological Control. (Ed.Yaxin Li), Cornell University, pp 75 –79.

Gavas, R., Kaushal, K.K. and Ganguly, S.(2004). Foraging behaviour and host range of an indigenous entomopathogenic nematode, *Steinernema thermophilum* Ganguly and Singh, 2000. In: *National Symposium on Paradigms in Nematological Research for Biodynamic Farming*, at UAS, Bangalore, Nov. 17-19, 2004; pp 122 –123.

Georgis, R. and Gaugler, R. (1991). Predictability in biological control using entomopathogenic nematodes. *J. Econ. Entomol.* 84: 713-720.

Georgis, R. (2002). The Biosys Experiment: an insider's perspective. In: *Entomopathogenic Nematology* (Ed. R. Gaugler), CABI Publishing, UK, pp 357–371.

Grewal, P.S., Gaugler, R. and Wang, Y. (1994). Enhanced cold tolerance of the entomopathogenic nematode *Steinernema feltiae* through genetic selection. *Ann. Appl. Biol.* 129: 335 –341.

Grewal, P.S., Lewis, E.E. and Gaugler, R. (1997). Response of infective stage parasites (Nematoda: Steinernematidae) to volatile cues from infected hosts. *Journal of Chemical Ecology* 23: 503-515.

Hunter, C. D. (1994). Suppliers of Beneficial Organisms in North America. California Environmental Protection Agency, Dept. Pesticide Regulation, *Environmental Monitoring and Pest Management Branch*. PM 94-03. pp 30.

Kaya, H.K. (1993). Entomogenous and Entomopathogenic nematodes in biological control. In: *Plant parasitic nematodes in temperate agriculture* (Eds. K. Evans, D.L. Trudgill and J.M. Webster), CAB International, Wallingford, pp 565-591.

Kaya, H.K. and Gaugler, R. (1993). Entomopathogenic nematodes. *Ann. Rev. Entomol.* 38: 181-206.

Koppenhofer, A.M., Grewal, P.S. and Kaya, H.K. (2000). Synergism of imidacloprid and entomopathogenic nematodes against white grubs: The mechanism. *Entomol. Exp. Appl.* 94: 283–293.

Lewis, E.E., Campbell, J.F. and Gaugler, R. (1998). A conservation approach for using entomopathogenic nematodes in turf and landscapes (II). In: *Perspectives on the Conservation of Natural Enemies of Pest Species* (Ed. P. Barbosa), Academic press, San Diego, pp 254-270.

Lewis, E.E. (2002). Behavioural Ecology. In: *Entomopathogenic Nematology* (Ed. R. Gaugler), CABI Publishing, UK, pp 205-224.

Lisansky, S.G. and Coombs, J. (1994). Developments in the market for biopesticides. Brighton crop protection conference - Pests and Diseases. pp 1049-1054.

McInerney, B.V., Gregson, R.P., Lacey, M.J., Akhurst, R.J., Lyons, G.R., Rhodes, S.H., Smith, D.R.J., Engelhardt, L.M. and White A.H. (1991a). Biologically active metabolites from *Xenorhabdus* spp. Part1. Dithiolopyrrolone derivatives with antibiotic activity. *J. Nat. Prod.* 54: 774-784.

Millar, L.C. and Barbercheck, M.E. (2001). Interaction between endemic and introduced entomopathogenic nematodes in conventional-till and no-till corn. *Biological Control.* 22: 235-245.

Millar, L.C. and Barbercheck, M.E. (2002). Effects of tillage practices on entomopathogenic nematodes in a corn agroecosystem. *Biological control* 25: 1-11.

Miller, G.T. (2004). *Sustaining the Earth*, 6th edition. Thompson Learning. Inc. Pacific Grove, California. Chapter 9: 211-216.

Minchella, D.J. and Scott, M.E. (1991). Parasitism-A cryptic determinant of animal community structure. *Trends in Ecology and Evolution* 6: 250-254.

Mukhopadhyay, A.N. (1994). Biocontrol of soil-borne fungal pathogen-current status, future prospects and potential limitations. *Indian Phytopathol.* 47: 1-8.

Nichols, A.E., Birch, M.R., Walter and Lindda, E. (1993). Plant products to control plant parasitic nematodes. *Pesticides Sci.* 39: 141-145.

Park, Y. and Kim, Y. (2000). Eicosanoids rescue *Spodoptera exigua* infected with *Xenorhabdus nematophilus*, the symbiotic bacteria to the entomopathogenic nematode, *Steinernema carpocapsae*. *J. Insect Physiol.* 46: 1469-1476.

Park, Y., Kim, Y., Putnam, S.M. and Stanley, D.W. (2003). The bacterium *Xenorhabdus nematophilus* depresses nodulation reactions to infection by inhibiting eicosanoid biosynthesis in tobacco hornworms, *Manduca sexta*. *Arch. Insect Biochem. Physiol.* 52: 71-80.

Paul, V.J., Frautschy, S., Fenical, W. and Nealson, K.H. (1981). Antibiotics in microbial ecology, isolation and structure assignment of several new antibacterial compounds from the insect-symbiotic bacteria *Xenorhabdus* spp. *J. Chem. Ecol.*, 7: 589-597.

Poinar, G.O. (1966). The presence of *Achromobacter nematophilus* Poinar and Thomas in the infective stage of a *Neoplectana* sp. (Steinermatidae: Nematoda). *Nematologica* 12: 105-108.

Poinar, Jr. G. O. (1983). *The natural history of nematodes*. Prentice hall, Inc., New Jersey. pp 323.

Poinar, G.O. (1990). Taxonomy and biology of Steinernematidae and Heterorhabditidae. In: *Entomopathogenic nematodes in biological control* (Eds. R. Gaugler and H.K. Kaya), CRC Press, Boca Raton, Florida, USA, pp 23-61.

Poinar, G.O. (1993). Origins and phylogenetic relationships of the entomophilic rhabditis, Heterorhabditis and Steinernema". *Fundamental and Applied Nematology* 16 : 333-338.

Poinar, G.O. Jr., Karunakar, G.K. and David, H. (1992). *Heterorhabditis indicus* n. sp. (Rhabditida, Nematoda) from India: separation of *Heterorhabditis* spp. by infective juveniles. *Fund. Appl. Nematol.* 15: 467-472.

Prasad, J.S., Sankar, M., Padmakumari, A.P. and Gururaj, K. (2009). Compatibility of entomopathogenic nematode, *Heterorhabditis Indica* with insecticides used in rice. In: International symposium on biopesticides, TERI, New Delhi, pp 55-53.

Rahman, P.F., Sharma, S.B. and Wightman, J.F. (2000). A review of insect-parasitic nematodes research in India: 1927-1997. *Int. J. Pest Manage.* 1: 19-28.

Rao, V.P. and Manjunath, T.M. (1966). DD-136 nematode that can kill many pests. *Indian Farming* 16: 43-44.

Sankar, M. (2009). Investigation of an indigenous entomopathogenic nematode, *Heterorhabditis indica* Poinar, Karunakar and David (1992) as a potential biocontrol agent of insect pest in rice. Ph.D. thesis, Osmania University, Hyderabad, pp. 168-79.

Shapiro-Ilan, D.I., Gough, D.H., Piggott, S.J. and Fife, J.P. (2006). Application technology and environmental considerations for use of entomopathogenic nematodes in biological control. *Biol. Control* 38: 124–133.

Strong, D. R., Kaya, H.K., Whipple, A.V., Child, A.L., Kraig, S. Bondonno, M., Dyer, K. and Maron, J.L. (1996). Entomopathogenic nematodes: natural enemies of root-feeding caterpillars on bush lupine. *Oecologia* (Berlin) 108: 167-173.

Sundar, L. and Chang, F.N. (1993). Antimicrobial activity and biosynthesis of indole antibiotics produced by *Xenorhabdus nematophilus*. *J. Gen. Microbiol.*, 129: 3139-3149.

Walter A., Geoffrey M., Jerome M., Louise N., Jennifer Sievert, Maria Propeck, Dorothy S., Alan Becker, Michelle Lackovic, Shannon B. S., Rupali Das, John B., Dorilee P, Catherine L., Thomsen, M.P.H., Martha Stanbury, (2005). Acute Illnesses Associated With Pesticide Exposure at Schools. *Journal of the American Medical Association* 294: 455–465.

Webster, J.M., Chen, G. and Li, J. (1998). Parasitic worms: An ally in the wax against the super bugs. *Parasitol. Today* 14: 161-163.

Webster, J.M., Chen, G., Hu, K. and Li, J. (2002). Bacterial metabolites. In: *Entomopathogenic Nematology*, Edited by R. Gaugler. CABI Publishing, New York. pp. 99-114.

Williams, R.J. and Heymann, D.L. (1998). Containment of antibiotic resistance. *Science*. 279:1153-1154.

Witte, W. (1998). Medical consequences of antibiotic use in agriculture. *Science*. 279: 996-997.

Wodageneh, A. and Wulp, H.V.D. (1997). Obsolete pesticides in developing countries. *Pesticides Information* 23: 33-36.

In: Advances in Biotechnology
Editors: H. K. Dhingra, P. N. Jha et al.
ISBN: 978-1-62417-842-9

Chapter 6

AERO-MICROBIOLOGY

Monika Sharma[1], Harish Dhingra[2,*] and Pratima Bajpai[2]
[1]Faculty of Arts, Science and Commerce,
[2]Mody Institute of Technology and Science, Laxmangarh, Rajasthan, India

ABSTRACT

Aeromicrobiology is the branch of science that deals with the study of micro-organism present in the air. Bioaerosols are usually defined as particulate matter of microbial, plant or animal origin that is often used equivalently with organic dust. Earth's atmosphere is known to team with airborne microorganisms, though the high light intensities, extreme temperature variations, low concentrations of organic matter, and a scarcity of water, make the atmosphere as unsuitable environment habitat for microbial growth. Four routes for the spread of microorganisms exist: contact, airborne, common vehicle, and vector-borne. Aeromicrobiology pathway includes three steps launching, transport and deposition of bioaerosols. Control of bioaerosol can be done by ventilation filtration and by using biocidal agents. Bioaerosol assessment is required in some industrial and environmental situations to assess the exposure of employees to airborne microorganisms such as bacteria, fungi or microbial toxins.

* Corresponding Author: E-mail:harishdhingra2000@gmail.com.

Keywords: Aeromicrobiology, bioaerosol, HEPA filter, ventilation

INTRODUCTION

In the 1970's a new era of Microbiology emerged and developed into the field of Environmental Microbiology. It is closely related to Microbial Ecology that is the study of the interaction of microorganisms within the environment. Primary difference between these two fields is that Environmental Microbiology is an applied field where the study of microorganisms in the environment leading to the benefit of the society is emphasized [1]. Aerobiology is a scientific discipline that deals with the transport of organisms and biologically significant materials through the atmosphere [2]. Aerobiology also include translocation, generation, uptake, dispersion, viability, deposition and nfection/infestation of seeds, viruses, fungi, bacteria and other agents, including insects such as aphids and mosquitoes, which act as virus vectors. Intramural Aerobiology deals with indoor environment while Extramural Aerobiology deals with outdoor environment.

Aeromicrobiology is the branch of science that deals with the study of micro-organism present in the air. Bioaerosols are usually defined as particulate matter of microbial, plant or animal origin that is often used equivalently with organic dust.

Bioaerosols or organic dust may consist of pathogenic or non-pathogenic live or dead bacteria and fungi, viruses, high molecular weight allergens, bacterial endotoxins, mycotoxins, peptidoglycans, pollen, plant fibres, etc [3]. Particles of biological origin usually vary in size from below 1 μm to approximately 50 μm or larger. Viruses typically range from 0.005 to 0.05 μm, while bacterial cells and spores typically range from 0.2 to 30 μm in length. Pollens and plant spores generally have more diameters between 10 and 100 μm. Assuming a density of 1.0 g/cm 3, the settling rate of a 0.1 μm diameter sphere in still air is 0.3 cm/h, a 1.0 μm sphere is 13 cm/h, and a 40 μm sphere is 300 cm/min, according to Stokes law [4]. Large airborne particles stay airborne for only short period Submicron particles will stay airborne for days. These are removed by sedimentation by wind or physical disturbance, by rain, diffusion to surfaces or by coagulation with other particles [5].

In this paper we will give an overview of the health effects associated with bioaerosol exposure. In addition, we will review exposure assessment methods.

MICROBIAL DIVERSITY IN AIR

Earth's atmosphere is known to team with airborne microorganisms, though the high light intensities, extreme temperature variations, low concentrations of organic matter, and a scarcity of water, make the atmosphere as unsuitable environment habitat for microbial growth [6].

Besides if a number of microorganism present in air those are introduce via agricultural practices, healthcare units and industrial operations such as sewage treatment, animal rendering, fermentation processes, and food processing plants [7]. Among the microorganisms present in the atmosphere, bacteria are often the highest in number, despite their high death rate due to environmental factors producing damage [8].

PATHOGENIC AIRBORN BACTERIA [9]

Baceria	**Disease**
Mycobacterium abscessus	Sporotrichoiddermatosis
Staphylococcus epidermidis	Septicemia and endocarditis
Bacillus subtilis	Dysentry
Pseudomonas aeruginosa	Pneumonia
Pseudomonas cepacia	Pneumonia and Bacteremia
Legionella pneumophila	Haemodialysis fistulae in respiratory tract
Mycobacterium chelonae	Obliterative bronchiolitis

PATHOGENIC AIRBORNFUNGI [10]

Fungi	**Disease**
Aspergillus	Aspergillosis
Absidia	-
Rhizopus	-
Mucor	-
Cryptococcus neoformans	Cryptococcosis
Histoplasmacapsulatum	Histoplasmosis
Blastomycesdermatitidis	Blastomycosis

Fungi	**Disease**
Coccidioidesimmitis	Coccidioidomycosis
Penicillium sp.	Respiratory irritation, allergic reactions
Micromonosporafaeni	Respiratory irritation, allergic reactions
Thermoactinomyces vulgaris	Respiratory irritation, allergic reactions
Alternaria	Respiratory irritation
Cladosporium	Respiratory irritation
Helminthosporium	Respiratory irritation

PATHOGENIC AIRBORNPROTOZOA [11]

Protozoa	**Disease**
Bodorepens	Pediatricdirofilariasis
Cercobodoradiatus	Gray crab disease
Colpodosteini	-
Mastigamoeba	-
Tetramitus	-
Monas	-

SIZE OF BIOAEROSOL

Studies have reported associations between particulate matter (PM), especially its PM10 fraction (size 10 μm), and health, even at air pollution levels below the National Ambient Air Quality Standard (NAAQS). Both respiratory and cardiovascular adverse effects have been reported, including premature mortality, respiratory and cardiovascular disease, decreased lung function, and increased risk of myocardial infarction [12-16]. The size of bioaerosol particles ranges between 0.002 and 100 μm (Whitby and Svendrup, 1980). Particles mainly consist of a mixture of a large number of chemical compounds, including sulfate, nitrate, ammonium ions, sea salt, organic and inorganic carbon, metals, and biological material. Size is an important determinant of particle fate and transport in the environment. It also affects deposition in the human respiratory system.

Coarse particles: The coarse fraction of PM10has a size range between 2.5 and 10 μm. These particles are typically generated mechanically by crushing or grinding processes. Coarse particles differ from fine particles (size 2.5 µm) not only because they are larger in size but also because they have different formation mechanisms, chemical composition, sources, and physical nature. Sources of coarse particles include resuspended dust, crustal material from roads and construction, mining, and farming activities, as well as bioaerosols, such as spores and pollen grains. Because the composition of coarse particles differs from that of fine and ultrafine ones, it is worth investigating their effects separately. The toxicity of coarse particles has been demonstrated in a limited number of *in vitro* studies [17]. Fine particles: The fine fraction of particulate matter has a size range between 0.1 and 1 µm. Fine particles sub–devided into two types: Nuclei mode (0.01-0.1 µm) and Accumulation mode (0.1-1.0 µm). It mainly content viruses.

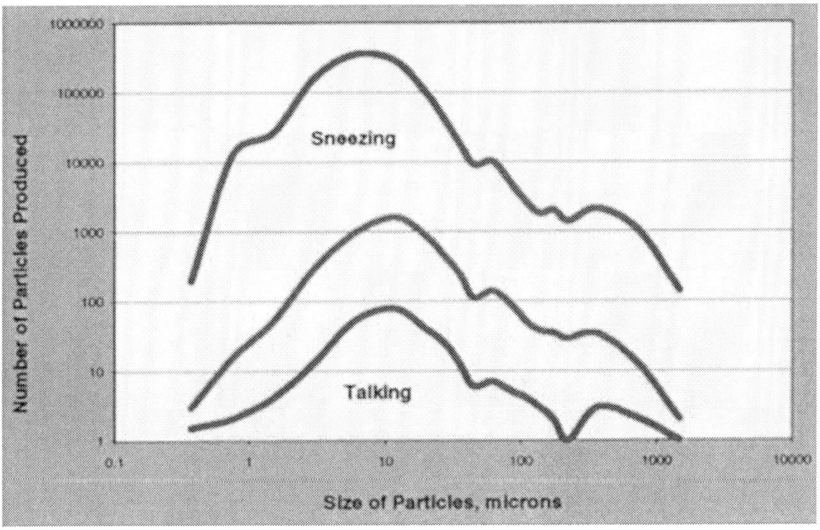

Figure. Profile of particle sizes produced by an infectious person. Based on data from Duguid, J. P., 1945 [18].

AIRBORN TRANSMISSION OF MICRO-ORGANISM

Four routes for the spread of microorganisms exist: contact, airborne, common vehicle, and vector-borne [19].

Contact transmission: It involves direct contact in which body-to-body contact takes place, or indirect in which the susceptible person comes into contact with a contaminated intermediate host.

Airborn transmission: Large droplet transmission is judged a form of contact transmission in which large droplets (>5 microns) contaminated with microorganisms are generated when an infected person sneezes, coughs, or talks. These droplets are propelled short distances (<1 m) and deposited on a susceptible host's conjunctiva or mucosa. Airborne transmission happens by aerolisation of an infectious agent through droplet nuclei (residua of large droplets containing microorganisms that have evaporated to <5 microns). These residual droplets become aerosolised and disperse widely, dependent on environmental conditions, and remain suspended in air for indefinite periods [20-22].

Common vehicle transmission: It involves one inanimate vehicle, which transmits infection to many hosts, and typically applies to microorganisms spread by food and water.

Vector-borne transmission: It involves the spread of disease by insects and vermin.

Successful spread of contagion to other hosts is dependent on many factors, including proximity, infectiousness of the source, pathogenicity of the microorganism, duration of exposure, environmental conditions (ventilation, humidity, temperature), and host-specific factors such as general health and immune status [23].

Aeromicrobiology Pathway

It includes three steps:

1. Launching of bioaerosol in air: That will make suspension in the air. The force responsible for suspension is air turbulence created by the any animal, human, machines. It may be action of either wind or water on the contaminated surface.
2. Transport of bioaerosol: Process of movement of bioaerosol from one place to the other. Based on theredistance travelled and the time taken in travelling, the way of transportation has been classified into the following types:
    a. Submicroscale transport: Movement of bioaerosol from one place to another upto 100 m within 10 min.
    b. Microscale transport: Transport of bioaerosol from a distance of 100m to 1km within 1hr.

c. Mesoscale transport: Bioaerosol moving a distance from 1km to 100km within 1hr to few days.

3. Deposition: Eventualy the bioaerosols will leave the air turbulence or turbulence of suspending gas. Because of the forces like gravitational force, downward diffusion and electrostatic deposition. Altimately the microorganism deposite on the surface.

CONTROL OF SPREAD OF BIOAEROSOL

Exposure to inhalable airborne particles and microbial agents may cause various health problems. Numerous techniques have been developed to reduce the exposure to indoor particles. Aerosol control in confined, poorly ventilated spaces, when the air exchange with filtration cannot be successfully applied, represents a particular challenge. Another challenge is to decrease the indoor concentration of specific airborne contaminants, e. g., viable biological particles like bacteria fungi and viruses. While some indoor air purification techniques aim solely at the aerosol concentration reduction, others are designed to inactivate viable bioaerosols (e. g., viruses, bacteria, protozoa and fungi) [24]. Some commercial air cleaners generate excessive ozone (as a primary biocidalagent); these devices have raised public health concerns [25]. Among various guidelines for ozone exposures, the following thresholds have been specified for occupational environments: 0.2 ppm for 2 h [26], 0.05-0.10 for 8 h [26], 0.1 ppm for 8 h [27], and 0.05 ppm for instantaneous (no time limit specified) exposure [28]. For comparison, the outdoor air standard is 0.08 ppm for 8 h [29]. Ozone generators can inactivate viable microorganisms; however, the inactivation occurs at concentrations significantly exceeding health standards [30, 31]. Photooxidation involving UV radiation and TiO2 as a photocatalyst has been applied for gas-phase detoxification of organic contaminants [32, 33] and for inactivating microorganisms in water [34, 35]. Some effort has been made to explore its application for air cleaning inside a closed-loop system [36, 37]. The investigators reported significant photocatalytic inactivation of stress-resistant Serratiamarcesens that occurred when aerosolized bacteria circulated in a closed-loop duct equipped with a TiO2 filter for a relatively long period of time [38], found similar effect for Escherichia coli, Microbacterium sp., and Bacillus subtilis; [39], reported considerable inactivation of airborne E. coli passing through a photoreactor coated with TiO2 film. The biocidal effect of the photocatalytic oxidation can be attributed to photogenerated valence-band holes, hydroxyl radicals, hydrogen peroxide, and other reactive oxygen species. Lin and Li [40], tested the viability change in airborne bacteria and fungi exposed to photooxidation inside a small photoreactor for a very short time, on the order of a second. No significant decrease in the colony forming unit (CFU) count was observed during such a short time. Furthermore, for hybrid air purifiers, which involve several air cleaning mechanisms, no sufficient information is available to differentiate their particle removal efficiency and the biocidal capabilities, which both aim at reducing the bioaerosol exposure in indoor air.

Ventilation

Spores form the most important group of non-communicable diseases. Outdoor spore levels vary with season and climate and can reach very high levels when dry, windy conditions result in disturbance of the soil where fungi grow. Surprisingly, few cases of respiratory infections have ever been attributed to inhalation of outdoor air, [41, 42]. Indoor air spore levels can differ from outdoor air in both concentration and composition of spores. In normal, dry buildings, spore levels tend to be anywhere from 10 to 100 percent of outdoor spore levels11 and are mostly less than 200 colony forming units (CFU) per cu meter. Problemfree, multi-story office buildings typically have levels that are 10 to 31 percent of the outdoor airlevels [43].

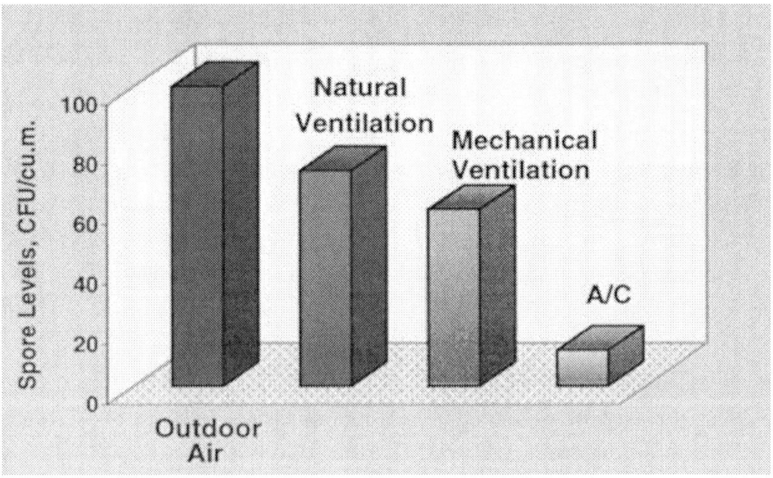

Figure. Indoor spore levels by ventilation system type. From theCalifornia Healthy Buildings Study [43].

So to reduce the spore concentration The Centers for Disease Control and Prevention (CDC) specifies a combination of administrative, engineering, and personal respiratory protection measures to achieve infection control [44].

Engineering controls include direct source control using local exhaust ventilation.

Flow of air in the contaminated room is known as ventilation. It is the most comman method being used for controlling the contamination of microorganism. The wind will naturally ventilate our home by entering or leaving windows, depending on their orientation to the wind. When wind blows against our home, air is forced into our windows.

When we know how air moves naturally through home or a contaminated chamber, we can then optimize it through mechanical ventilation. Mechanical ventilation can provide continuously moving air that will keep contaminated chamber aeriated, day and night, with circulating fans, whole-house fans and airconditionars. The quality and energy efficiency of these devices varies widely. Keeping a clear path for airflow both inside and outside your windows that is the main motiv of ventilation.

Filtration

Filtration is the separation of solid particles from a gaseous (Air) or Liquid suspension by passing the suspension (Air or Liquid) through a septum (or membrane) that retains most of the solids on or within itself. The septum (or membrane) is called a filter medium. The equipment that holds the filter medium and sprovides space for the accumulated solids is called a filter. Bioaerosols residing on filters generate rank odors and could be re-suspended in the air due to reverse airflow caused by a temporary reversal of the surrounding pressure or breakdown and maintenance of the filters. Therefore, there is a demand for the development of new filter materials that are capable of addressing these problems regarding deposited bioaerosols. Filters treated with anti-microbial components such as iodine [45, 46], and membrane-breaking enzymes have been developed and tested as anti-microbial filters. However, when non-biological dust is deposited on the surfaces of filters during operation, the attached antimicrobial materials will be instantly covered by dust and their effects nullified. Therefore, in general, anti-microbial filters have short effective times due to the accumulation of non-biological dust from the air. Therefore, recent developments include hybrid methods that combine several different bioaerosol control mechanisms.

Biocidal Agents

Super heating, super dehydration, ozonization and UV radiation can be used as an effective method for control of bioaerosols. UV radiation that is a germicidal agent that can be used in rooms and hotels. Ultraviolet (UV) irradiation is a widely used method for controlling bioaerosols in indoor environments. The purine and pyrimidine bases of DNA (deoxyribonucleic acid) strongly absorb UV irradiation that is between 220 and 300 nm in wavelength, resulting in modifications or breaks. This sometimes leads to disruption of the genetic function and the death of the exposed organisms [47]. One study demonstrated a 12-fold reduction in bacterial colony formation in an operating room upon exposure to a high level of UVGI at 290 μW/cm2 [48]. The artificial emission of air ions (ion density of 105–106 e±cm-3) for 30 minutes results in the removal of 97% of 0.1 μm particles and 95% of 1 μm particles from indoor air in addition to the natural decay effect (naturally occurring decrease in aerosol concentration due to gravity and diffusion) [49].

Isolation Chambers

Certain type of isolation chambers can be prepared for caontrol of bioaerosols. These isolation chambers can be two types:

1. Negatively pressurised isolation chamber: In this type of isolation chamber air is sucked from room then it passed through the HEPA filter then transferred from UV radiation biocidal agent chamber then released into atmosphere.example:TB ward in hospitals.
2. Positively pressurised isolation chamber: In rest of the hospital positively pressurised isolation chamber is maintain. It work in the opposite principle than Negatively pressurised isolation chamber. Pumping of the air into the outside the enviroment. So the occupment of rest of the air by outside enviroment.

Assessment of the Air Quality

Bioaerosol assessment is required in some industrial and environmental situations to assess the exposure of employees to airborne microorganisms such as bacteria, fungi or microbial toxins. To measure bioaerosol concentrations, new sampling techniques have recently been developed either from existing dust samplers [50] or from original designs [51]. When sampling microorganisms, the viability and the reproductive activity of the biological agents should be maintained. They should be easily recoverable from the sampling substrate for culture and for analytical purposes. At least, the interval between the limit of detection and the limit of saturation of both, namely the sampling and analytical methods, should cover the expected bioaerosol concentration.

In general, there are many kinds of principles applied to bioaerosol sampling in order to collect and cultivate microorganisms: impaction of the particles onto nutrient agar, particle filtration by membrane filters or particle sampling into a liquid (e. g., bubblers, impingers). The agar impactors are quickly saturated in the case of higher microbiological concentrations. Desiccation of microorganisms can occur during filter sampling. Sampling microorganisms directly into a Liquid provides suitable samples for biological analysis, but bubblers often have a low physical sampling efficiency and impingers can stress the microorganisms [52, 53]. Other methods of sampling directly into liquid are available, [54] e. g., wet cyclones or Venturi tubes. Collection of bioaerosol can occur by using the internal force of gravity and because the size of exclusion. Collection of their born particles can possible by using the certain natural deposition forces like gravitation force and other downward forces. There is still a considerable need to develop new bioaerosol sampling methods, mainly for purposes of personal sampling.

Conclusion

This article reviews the presence of microorganisms in air and their sources, the relation of airborne dust and endotoxin, the sources of atmospheric microbial contamination in food-processing plants, the mechanisms of airborne particle deposition, the importance of airborne microbes, the survival of microorganisms in air, methods of air sampling, airborne microbial populations in food-processing plants, control of airborne microorganisms in food-processing plants, and the general issue of microorganisms in air and their impact on food safety. The purpose was to bring together scattered information about airborne microorganisms and review their importance in food protection and sanitation.

REFERENCES

[1] Atlas, R. M., 1997. *Principles of Microbiology*, 2nd edition. Wm. C. Brown Publishers, USA.

[2] Isard, S. A. and Gage, S. H., 2001. Flow of Life in the Atmosphere: *An Airscape Approach to Understanding Invasive Organisms.* East Lansing, Michigan State University Press.

[3] van Tongeren, M., van Amelsfoort L. and Heederik D., 1997. Exposure to organic dusts, endotoxins, and microorganisms in the municipal waste industry. *Int J Ind Med*; 33:384–91.

[4] Corn, M., 1968. *Non viable particles in the air*, in Air Pollution, 2nd edition. Ed. Stern A. C., Academic Press, New York, USA. 47-94.

[5] Cox, C. S. and Wathes, C. M., 1995. *Bioaerosols Handbook*, CRC Press, ISBN 1-87371-615, 285-316.

[6] Atlas, R. M., 1984. *Microbiology, Fundamentals and Applications* Macmillan Publishing Co., New York, London.

[7] Cullinan, P., Cook, A. and Nieuwenhuijsen, M. J., 2001. Allergen and dust exposure as determinants of work related symptoms and sensitization in a cohort of flour exposed workers; a case-control analysis, *Annals Occupational Hygiene* 45: 97-103.

[8] Madrioli, P., Comtois, P. and Levizzani, V., 1998. *Methods in aerobiology* Pitagova 5th edition, Bologna, Italy, 262.

[9] McCullough N. V., Brosseau, L. M. and Vesley, D., 1997. Collection of three bacterial aerosols by respirator and surgical mask filters under varying conditions of flow and relative humidity, *The Annals of Occupational Hygiene,* 41 (6)677-690.

[10] Burge, H., 1997. Fungi: How they grow and their effects on human health. *HPAC*, 69(6).

[11] Rivera, F., Lugo, A., Ramirez, E., Bonilla, P., Calderon, A., Rodriguez, S., Ortiz, R., Gallegos, E., Labastida, A., and Chave, M. P., 1991. Seasonal distribution of air-borne protozoa in mexico city and its suburbs, *Springerlink:journal of earth and enviromental science*, 61:17-36.

[12] Schwartz, J., and Dockery, D. W., 1992. Increased mortality in Philadelphia associated with daily air pollution concentrations, *Am. J. Epidemiol.* 135:12–19.

[13] Dockery, D. W., Pope, C. A. III, Xu, X., Spengler, J. D., Ware, J. H., Fay, M. E., Ferris, B. G., and Speizer, F. E., 1993. An association

between air pollution and mortality in six U. S. Cities, *N. Engl. J. Med.* 329:1753–1808.

[14] Dockery, D. W., and Pope, C. A., 1994. Acute effects of particulate air pollution, *Annu. Rev. Pub. Health* 15:107–132.

[15] Peters, A., Dockery, D. W., Heinrich, J., and Wichman, H. E., 1997. Short-term effects of particulate air pollution on respiratory morbidity in asthmatic children, *Eur. Respir. J.* 10:872–879.

[16] Gamble, J., 1998. PM2.5 and mortality in Long term prospective cohort studies: Cause effects or statistical associations? *Environ. Health Perspect.* 106(9):535–549.

[17] Monn, C., and Becker, S., 1998. Fine and coarse particles: Induction of cytokines in human monocytes, *J. Aerosol Sci.* 29:305–306.

[18] Duguid, J. P., 1945. The size and the duration of air-carriage of respiratory droplets and dropletnuclei, *Journal of Hygiene* 54: 471-479.

[19] Chin, J., 2000. *Control of communicable diseases manual*, Washington, DC: American Public Health Association.16.

[20] Withers, M. and Christopher, G., 2000. Aeromedical evacuation of biological warfare casualties: a treatise on infectious diseases on aircraft. *Mil Med*, 165 (3): 1–21.

[21] Nardell, E., Keegan, J., Cheney, S. and Etkind, S. C., 1991. Airborne infection: theoretical limits of protection achievable by building ventilation, *Am Rev Respir Dis*, 144: 302–06.

[22] Riley, R. and Nardell, E., 1989. Clearing the air, *Am Rev Respir Dis*, 139:1286–94.

[23] Nagda, N. L., Fortmann, R. C., Loontz, M. D., Baker, S. R, and Ginevan, M. E., 1989. *Airliner cabin environment: contaminant measurements, health risks, and mitigation options*, Washington, DC: US Department of Transportation. 15:89-5.

[24] Grinshpun, S. A., Willeke, K., Ulevicius, V., Juozaitis, A., Terzieva, S., Donelly, J., Stelma, G. N. and Brenner, K. P., 1997. Effect of impaction bounce and reaerosolization on the collection efficiency of impingers, *Aerosol Sci. Technol.*, 26:326–342.

[25] EPA (U. S. Environmental Protection Agency). Ozone generators that are sold as air cleaners: *An assessment of effectiveness an health consequences;* http://www.epa.gov/iaq/pubs/ozonegen.html; Accessed on May 22, 2006.

[26] *ACGIH (American Conference of Governmental Industrial Hygienists).* Threshold Limit Values for Chemical Substances and Physical Agents & Biological Exposure Indices; ACGIH:Cincinnati, OH, 2004; 44.

[27] OSHA (Occupational Safety & Health Administration). Regulations (Standards 29 CFR): *Air Contaminants,* 1915.1000; http://www.osha.gov/pls/oshaweb; Accessed on May 31, 2006.

[28] FDA (U. S. Food and Drug Administration). *Labeling Regulatory* Requirements for Medical Devices; HHS publication FDA 89-4203; U. S. Governement Printing Office: Washington, DC, 1989.

[29] EPA (U. S. Environmental Protection Agency). *National Ambient Air Quality Standards* (1997); http://www.epa.gov/air/criteria.html; Accessed on May 24, 2006.

[30] Foarde, K. K., van Osdell, D. W. and Steiber, R. S., 1997. Investigation of gas-phase ozone as a potential biocide, *Appl. Occup. Environ. Hyg.* 12:535-542.

[31] Li, C. S. and Wang, Y. C., 2003. Surface germicidal effects of ozone for microorganisms. *AIHA J.* 64: 533-537.

[32] Alberici, R. M., Canela, M. C., Eberlin, M. N. and Jardim, W. F., 2000. Catalyst deactivation in the gas phase destruction of nitrogen-containing organic compounds using TiO2/UV-VIS. *Appl. Catal.,* 793: 1-9.

[33] Shang, J., Du, Y.and Xu, Z., 2002. Photocatalytic oxidation of heptane in the gas-phase over TiO2. *Chemosphere*, 46:93-99.

[34] Matsunaga, T., Tomoda, R., Nakajima, T., Nakamura, N. and Komine, T., 1988. Continuous sterilization system that uses photosemiconductor powders. *Appl. Environ. Microbiol,* 54:1330-1333.

[35] Sunada, K., Kikuchi, Y., Hashimoto, K. and Fujishima, A., 1998. Bactericidal and detoxification effects of TiO2 thin film photocatalysts. *Environ. Sci. Technol,* 32:726-728.

[36] Goswami, D. Y., Trivedi, D. M. and Block, S. S., 1997. Photocatalytic disinfection of indoor air. *Trans. ASME* 119: 92-96.

[37] Goswami, T. K., Hingorani, S.,Greist, H., Goswami, D. Y.and Block, S. S., 1999. Photocatalytic system to destroy bioaerosols in air. *J Adv. Oxidation Tech,* 4:185-188.

[38] Pal, A., Min, X., Yu, L. E., Pehkonen, S. O. and Ray, M. B., 2005. Photocatalytic inactivation of bioaerosols by TiO2 coated membrane. *Int. J. Chem. Reactor Eng*, 3: A45.

[39] Keller, V., Keller, N., Ledoux, M. J. and Lett, M. C., 2005. Biological agent inactivation in a flowing air stream by photocatalysis. *Chem. Commun.* 23:2918-2920.

[40] Lin, C. Y. and Li, C. S., 2003. Effectiveness of titanium dioxide photocatalyst filters for controlling bioaerosols. *Aerosol Sci. Technol*, 37:162-170.

[41] Ryan, K. J. ed., 1994. *Sherris Medical Microbiology*, Norwalk: Appleton & Lange, 221-250.

[42] Mandell, G. L. ed., 1985. *Principles and Practice of Infectious Diseases.,* New York: Wiley.

[43] Godish, T., 1995. Sick Buildings: *Definition, Diagnosis, and Mitigation.* Boca Raton: Lewis Publishers.

[44] CDC, 1994. Guidelines for Preventing the Transmission of Mycobacterium tuberculosis in Healthcare Facilities. *MMWR* 43(RR-13): 1-132.

[45] Lee, J. H., Wu, C. Y., Wysocki, K. M., Farrah, S. and Wander, J., 2008b. Efficacy of Iodine Treated Biocidal Filter Media Against Bacterial Spore Aerosols, *J. Appl. Microbiol.* 105: 1318–1326.

[46] Eninger, R. M., Adhikari, A., Reponen, T. and Grinshpun, S. A., 2008. Differentiating between Physical and Viable Penetrations When Challenging Respirator Filters with Bioaerosols. *Clean* 36: 615–621.

[47] Madigan, M. T. and Martinko, J. M. 2006. *Brock Biology of Microorganisms.* 11th Edition, Pearson Prentice Hall, New Jersey, USA, Chap. 1, Chap 3 (p. 52), Chap 20 (p.673).

[48] Lidwell, O. M., 1994. Ultraviolet Radiation and the Control of Airborne Contamination in the Operating Room, *J. Hosp. Infect.* 28: 245–248.

[49] Lee, B. U., Yermakov, M. and Grinshpun, S. A., 2004b. Removal of Fine and Ultrafine Particles from Indoor Air Environments by the Unipolar Ion Emission, *Atmos. Environ.* 38: 4815–4823.

[50] Kenny, L. C., Stancliffe, J. D., Crook, B., Stagg, S., Griffiths, W. D., Stewart, I. W. and Futter, S. J., 1999. *Bacillus thuringiensis* in Fecal Samples from Greenhouse Workers after Exposure to B. thuringiensis-Based Pesticides, *Am. Ind. Hyg. Assoc. J.,* 59:831–841.

[51] Reponen, T., Willeke, K., Grinshpun, S. A. and Nevalainen, A., 2001. Aerosol Measurement - Principles, Techniques and Applications, ed. P. A. Baron and K. Willeke, *Wiley Interscience*, *Chichester,* 2nd ed, 751–777.

[52] Grinshpun, S. A., Willeke, K., Ulevicius, V., Juozaitis, A., Terzieva, S., Donelly, J., Stelma, G. N. and Brenner, K. P., 1997. Effect of impaction bounce and reaerosolization on the collection efficiency of impingers, *Aerosol Sci. Technol.*, 26:326–342.

[53] Lin, W. H. and Li, C. S., 1990.Optimization of a Sampling System for Recovery and Detection of Airborne Porcine Reproductive and Respiratory Syndrome Virus and Swine Influenza Virus *Aerosol Sci. Technol.*, 30:109–118.

[54] Licht, W., 1988. Air Pollution Control Engineering - *Basic calculations for particulate collection*, Marcel Dekker Inc., New York, USA, 2nd ed.

In: Advances in Biotechnology
Editors: H. K. Dhingra, P. N. Jha et al.
ISBN: 978-1-62417-842-9

Chapter 7

ADVENT OF REAL TIME PCR OVER CONVENTIONAL PCR

***Harish Dhingra*[1,*], *Pratima Bajpai*[2] *and Anamika Pandey*[3]**

[*1,2]Department of Science, FASC, MITS University, Lakshmangarh, Sikar, Rajasthan, India

[3]ACTREC,Tata memorial Centre, Kharghar, Navi Mumbai, India

ABSTRACT

Polymerase chain reaction (PCR) is a method that allows amplification of a gene or short DNA sequences within a longer double stranded DNA molecule. Polymerase Chain Reaction was invented by Kary B. Mullis in the mid-1980s & this technique revolutionized molecular biology. PCR is a fairly standard procedure now, and its use is extremely wide-ranging. PCR entails the use of a pair of primers, each about 20 nucleotides in length that are complementary to a defined sequence on each of the two strands of the DNA. These primers are extended by a DNA polymerase so that a copy is made of the designated sequence. After making this copy, the same primers can be used again, not only to make another copy of the input DNA strand but also of the short copy made in the first round of synthesis. This leads to exponential amplification. Since it is necessary to raise the temperature to separate the two strands of the double strand DNA in each round of the amplification

[*] Corresponding Author: e.mail: Harish Dhingra, E-mail: harishdhingra2000@gmail.com.

process, a major step forward was the discovery of a thermo-stable DNA polymerase (Taq polymerase) that was isolated from *Thermus aquaticus,* a bacterium that grows in hot pools; as a result it is not necessary to add new polymerase in every round of amplification. After several (often about 40) rounds of amplification, the PCR product is analyzed on an agarose gel and is abundant enough to be detected with an ethidium bromide stain. But this method of analysis of PCR product is semi-quantitative and, in many cases, the amount of product is not related to the amount of input DNA making this type of PCR a qualitative tool for detecting the presence or absence of a particular DNA. In order to measure messenger RNA (mRNA), the method was extended using reverse transcriptase to convert mRNA into complementary DNA (cDNA) which was then amplified by PCR and, again analyzed by agarose gel electrophoresis. In many cases this method has been used to measure the levels of a particular mRNA under different conditions but the method is actually even less quantitative than PCR of DNA because of the extra reverse transcriptase step. Reverse transcriptase-PCR analysis of mRNA is often referred to as "RT-PCR" which is unfortunate as it can be confused with "real time-PCR".

Keywords: PCR, Polymerase, DNA, RNA, probe

INTRODUCTION

A polymerase will synthesize a complementary sequence of bases to any single strand of DNA providing it has a double stranded starting point. This is very useful because you can choose which gene you wish the polymerase to amplify in a mixed DNA sample by adding small pieces of DNA complimentary to your gene of interest. These small pieces of DNA are known as primers because they prime the DNA sample ready for the polymerase to bind and begin copying the gene of interest.

During a PCR, changes in temperature are used to control the activity of the polymerase and the binding of primers. To begin the reaction the temperature is raised to 95°C. At this temperature all double stranded DNA is “melted” in to single strands. The temperature is then lowered to ~50°C.

This allows the primers to bind to your gene of interest. Thus the polymerase has somewhere to bind and can begin copying the DNA strand. The optimal temperature for the polymerase to operate is 72°C so at this point the temperature is sometimes raised to 72°C to allow the enzyme to work faster. There are now twice as many copies of your gene of interest as when you started. The cycle of changing temperatures (95°C, 50°C and 72°C) is then repeated and two copies become four. Another cycle and four become eight, and so on for 40-50 cycles.

After amplifying your gene in to many millions of copies it is possible to run the amplified DNA out on a polyacrylamide gel and stain it with a dye which makes is visible. The bigger the visible band, the more copies of your gene you have created.

Real-Time PCR

Real-Time PCR is identical to a simple PCR except that the progress of the reaction in monitored by a camera or detector in "real-time". There are a number of techniques that are used to allow the progress of a PCR to be monitored. Each technique uses some kind of fluorescent marker which binds to the DNA. Hence as the number of gene copies increases during the reaction so the fluorescence increases. This is advantageous because the efficiency and rate of the reaction can be seen. There is also no need to run the PCR product [1]out on a gel after the reaction.

Methods of Monitoring DNA Amplification in "Real-time"

Fluorescent Dyes

Intercalating fluorescent dyes (e.g. SYBR green) are the simplest and cheapest way to monitor a PCR in real-time. These dyes fluoresce only when bound to double-stranded DNA. So as the number of copies of DNA increases during the reaction so the fluorescence increases. The major disadvantage of using a dye such as this is the lack of specificity. This dye will report the amplification of any DNA not just your gene of interest.

Fluorescent Probes

Fluorescent probes are pieces of DNA complimentary to your gene of interest that are labeled with a fluorescent dye. The simplest and most commonly used type of probe is the Taqman-type probe. These probes are labeled with a fluorescent reporter molecule at one end and a quencher molecule (capable of quenching the fluorescence of the reporter) at the other. Hence under normal circumstances the fluorescent emission from the probe is low. However during the PCR the probe binds to the gene of interest and becomes cleaved by the polymerase. Hence the reporter and quencher are physically separated and the fluorescence increases [2]. Another commonly used type of probe is the "molecular beacon". Again these are small pieces of DNA complimentary to your gene of interest labeled with a fluorescent reporter and a quencher molecule on opposite ends. These probes are designed to fold on to themselves to bring the reporter and quencher in to closer proximity and minimize fluorescent emission. However, when the probe binds to the gene of interest the probe takes up a linear confirmation and the reporter and quencher are separated. This results in the desired increase in fluorescence. Molecular beacon probes are not cleaved by the polymerase but are simply "knocked off" again. The output from a real-time PCR reaction is in the form of a graph showing the number of PCR cycles (1 cycle = 90°C, 50°C, 72°C) against the increasing fluorescence [3,4].

The basic goal of real-time PCR is to precisely distinguish and measure specific nucleic acid sequences in a sample even if there is only a very small quantity. Real-time PCR amplifies a specific target sequence in a sample then monitors the amplification progress using fluorescent technology. During amplification, how quickly the fluorescent signal reaches a threshold level correlates with the amount of original target sequence, thereby enabling quantification. In addition, the final product can be further characterized by subjecting it to increasing temperatures to determine when the double-stranded product "melts." This melting point is a unique property dependent on product length and nucleotide composition [5]. To accomplish these tasks, conventional PCR has been coupled to state-of-the-art fluorescent chemistries and instrumentation to become real-time PCR.

At its core, real-time PCR technology utilizes conventional PCR. PCR is a procedure by which DNA can be copied and amplified. PCR exploits DNA polymerases to amplify specific pieces of DNA using short, sequence-specific oligonucleotides added to the reaction to act as primers. The first and most commonly used of these enzymes is *Taq* DNA polymerase (from *Thermus aquaticus*), whereas *Pfu* DNA polymerase (from *Pyrococcus furiosus*) is used widely because of its higher fidelity when copying DNA. Although these enzymes are subtly different, they both have two basic capabilities that make them useful for PCR: *1*) they can generate new strands of DNA using a DNA template and primers, and *2*) they are heat resistant. The latter attribute is necessary because after each round of DNA copying, the resulting double-stranded DNA (dsDNA) must be "melted" into single strands by high temperatures within the reaction tube (~95°C). The reaction is then cooled to allow the oligonucleotide primers to anneal to the now single-stranded template DNA and direct the DNA polymerase enzyme to initiate elongation by adding single complementary nucleotides to create a new complete strand of DNA. Thus dsDNA is created. This new dsDNA must then be melted apart before the next cycle of copying can occur. Therefore, if the reaction works with perfect efficiency, there will be twice as much specific dsDNA after each cycle of PCR. In reality, reactions do not maintain perfect efficiency because reactants within the PCR are consumed after many cycles, and the reaction will reach a plateau [6,7,8,9]. In addition, self-annealing of the accumulating product may also contribute to the "plateau effect". In fact, it is this very attribute of PCRs that makes real-time PCR technology so necessary. Because the reaction is able to efficiently amplify DNA only up to a certain quantity before the plateau effect, there is no way to reliably calculate the amount of starting DNA by quantifying the amount of product at the completion of the PCR. That is to say, no matter how much of a specific target DNA sequence is present before PCR, there can be similar amounts of amplified DNA after PCR, and any distinct correlation between starting and finishing quantities is lost. Real-time PCR addresses this problem by taking advantage of the fact that DNA amplifications do occur efficiently early in the reaction process and, therefore, measures product formation during this "exponential phase". This measurement correlates to the amount of specific starting DNA, thereby allowing quantification.

Reverse Transcriptase Extends the Utility of Real-time PCR

A major limitation of DNA polymerases (and PCR itself) is that they generally must use DNA as their template. They cannot, for example, amplify RNA in a similar manner. This problem can be overcome by another enzyme, reverse transcriptase, which generates complementary DNA (or cDNA) from an RNA template. Reverse transcriptases are enzymes used in nature by retroviruses, including human immunodeficiency virus and hepatitis C virus, to generate DNA from viral RNA. The virus derived DNA can then be inserted into the host's genome. In the laboratory, reverse transcriptase is used to convert RNA to cDNA that can then be used for multiple purposes including PCR-based applications. There are several commonly used reverse transcriptases, including avian myeloblastosis virus reverse transcriptase, Moloney murine leukemia virus reverse transcriptase, or engineered enzymes that enhance polymerase activity or decrease unwanted nuclease activities (e.g., Omniscript, PowerScript, StrataScript, SuperScript II, etc.). Under the appropriate reaction conditions, the relative amount of a given cDNA generated by reverse transcription is proportional to the relative amount of its RNA template. If this were not the case, measurements of cDNA quantities would have no relevance to RNA [10,11,12]. Reliably generated cDNA can be used as the raw material for real-time PCR, thereby utilizing its precision and sensitivity to determine changes in gene expression (i.e., RNA levels).

This is called real-time RT-PCR and has become the most popular method of quantitating steady-state mRNA levels. It is most often used for two reasons: either as a primary investigative tool to determine gene expression or as a secondary tool to validate the results of DNA microarrays. Because of the precision and sensitivity of real-time RT-PCR, even subtle changes in gene expression can be detected. Thus real-time PCR can be used to assess both DNA and RNA levels with great sensitivity and precision.

The Chemistries of Real-time PCR

The key to real-time PCR is the ability to monitor the progress of DNA amplification in real time. This is accomplished by specific chemistries and instrumentation. Generally, chemistries consist of special fluorescent probes in the PCR. Several types of probes exist, including DNA-binding dyes like EtBr or SYBR green I, hydrolysis probes (5'-nuclease probes), hybridization probes, molecular beacons, sunrise and scorpion primers, and peptide nucleic acid (PNA) light-up probes. Each type of probe has its own unique characteristics, but the strategy for each is simple. They must link a change in fluorescence to amplification of DNA [13,14,15,16].

SYBR green I bind to the minor groove of dsDNA, emitting 1,000-fold greater fluorescence than when it is free in solution. Therefore, the greater the amount of dsDNA presents in the reaction tube, the greater the amount of DNA binding and fluorescent signal from SYBR green I. Thus any amplification of DNA in the reaction tube is measured. Other dsDNA-specific dyes (e.g., BEBO, YOYO-1, TOTO-1, etc.) have also been described but are not as widely used. The primary concern with the usage of any of these sequence-independent dsDNA-binding probes is specificity. To help ensure specificity, the dissociation curve of the amplified product can be analyzed to determine the melting point. If there are two or more peaks, it suggests that more than one amplified sequence was obtained, and the amplification was not specific for a single DNA target [17,18,19,20].

Hydrolysis probes (also called 5'-nuclease probes because the 5'-exonuclease activity of DNA polymerase cleaves the probe) offer an alternative approach to the problem of specificity. These are likely the most widely used fluorogenic probe format and are exemplified by TaqMan probes. In terms of structure, hydrolysis probes are sequence-specific dually fluorophore-labeled DNA oligonucleotides. One fluorophore is termed the quencher and the other is the reporter. When the quencher and reporter are in close proximity, that is, they are both attached to the same short oligonucleotide, the quencher absorbs the signal from the reporter. This is an example of fluorescence resonance energy transfer (also called Förster transfer) in which energy is transferred from a "donor" (the reporter) to an "acceptor" (the quencher) fluorophore. During amplification, the oligonucleotide is broken apart by the action of DNA polymerase (5'-nuclease activity) and the reporter and quencher separate, allowing the reporter's energy and fluorescent signal to be liberated. Thus destruction or hydrolysis of the oligonucleotide results in an increase of reporter signal and corresponds with the specific amplification of DNA. Examples of common quencher fluorophores include TAMRA, DABCYL, and BHQ, whereas reporters are more numerous (e.g., FAM, VIC, NED, etc). Hydrolysis probes afford similar precision as SYBR green I, but they give greater insurance regarding the specificity because only sequence-specific amplification is measured. In addition, hydrolysis probes allow for simple identification of point mutations within the amplicon using melting curve analysis. There are several other variations on the reporter-quencher theme, including molecular beacons, sunrise primers, and scorpion primers. They each seek to keep the reporter and quencher together before amplification while separating them and generating the fluorescence signal during amplification [21,22,23,24,25]. Another class, called hybridization probes, uses donor and acceptor fluorophores, whereas PNAs containing thiazole orange fluorophores (called light-up probes) also emit greater signal upon binding of DNA. These do not represent an exhaustive list, as many other specific and nonspecific chemistries exist. In addition, new fluorescent chemistries are continually being developed with a focus on increasing sensitivity (by increasing the signal-to-noise ratio) and specificity, enhancing multiplexing capabilities, and reducing cost.

The Instrumentation of Real-time PCR

A critical requirement for real-time PCR technology is the ability to detect the fluorescent signal and record the progress of the PCR. Because fluorescent chemistries require both a specific input of energy for excitation and a detection of a particular emission wavelength, the instrumentation must be able to do both simultaneously and at the desired wavelengths.

Thus the chemistries and instrumentation are intimately linked. At present, there are three basic ways in which real-time instrumentation can supply the excitation energy for fluorophores: by lamp, light-emitting diode (LED), or laser. Lamps are classified as broad-spectrum emission devices, whereas LEDs and lasers are narrow spectrum. Instruments that utilize lamps (tungsten halogen or quartz tungsten halogen) may also include filters to restrict the emitted light to specific excitation wavelengths. Instruments using lamps include Applied Biosystem's ABI Prism 7000, Stratagene's Mx4000 and Mx3000P, and Bio-Rad's iCycler iQ. LED systems include Roche's LightCycler, Cepheid's SmartCycler, Corbett's Rotor-Gene, and MJ Research's DNA Engine Opticon 2. The ABI Prism 7900HT is the sole machine to use a laser for excitation [26,27,28,29].

To collect data, the emission energies must also be detected at the appropriate wavelengths. Detectors include charge-coupled device cameras, photomultiplier tubes, or other types of photodetectors. Narrow wavelength filters or channels are generally employed to allow only the desired wavelength(s) to pass to the photodetector to be measured. Usually, multiple discrete wavelengths can be measured at once, which allows for multiplexing, i.e., running multiple assays in a single reaction tube.

Another portion of the instrumentation consists of a thermocycler to carry out PCR. Of particular importance for real-time PCR is the ability of the thermocycler to maintain a consistent temperature among all sample wells, as any differences in temperature could lead to different PCR amplification efficiencies. This is accomplished by using a heating block (Peltier based or resistive), heated air, or a combination of the two. As one might expect, heating blocks generally change temperature more slowly than heated air, resulting in longer thermocycling times. For example, Roche's LightCycler models utilizing heated air can perform 40 cycles in 30 min, whereas Applied

Biosystem's ABI Prism 7900HT utilizing a Peltier-based heating block takes 1 h 45 min. For further details on real-time PCR instrumentation, resources include Real-Time PCR: an Essential Guide for a side-by-side comparison or the respective manufacturer's website for the most up to date model information.

Real-time instrumentation certainly would not be complete without appropriate computer hardware and data-acquisition and analysis software. Software platforms try to simplify analysis of real-time PCR data by offering graphical output of assay results including amplification and dissociation (melting point) curves. The amplification curve gives data regarding the kinetics of amplification of the target sequence, whereas the dissociation curve reveals the characteristics of the final amplified product [30].

Advantages and Limitations of Real-time PCR Quantitation

There are many methods in molecular biology for measuring quantities of target nucleic acid sequences. However, most of these methods exhibit one or more of the following shortcomings: they are time consuming, labor intensive, insufficiently sensitive, nonquantitative, require the use of radioactivity, or have a substantial probability of cross contamination. These methods include but are not limited to Northern and Southern hybridizations, HPLC, scintillation proximity assay, PCR-ELISA, RNase protection assay, in situ hybridization, and various gel electrophoresis PCR end-point systems. Real-time PCR has distinct advantages over these earlier methods for several reasons. Perhaps the most important is its ability to quantify nucleic acids over an extraordinarily wide dynamic range (at least 5 log units). This is coupled to extreme sensitivity, allowing the detection of less than five copies (perhaps only one copy in some cases) of a target sequence, making it possible to analyze small samples like clinical biopsies or miniscule lysates from laser capture microdissection. With appropriate internal standards and calculations, mean variation coefficients are 1–2%, allowing reproducible analysis of subtle gene expression changes even at low levels of expression. In addition, all real-time platforms are relatively quick, with some affording high-throughput automation [31].

Finally, real-time PCR is performed in a closed reaction vessel that requires no post-PCR manipulations, thereby minimizing the chances for cross contamination in the laboratory. However, there are several limitations to real-time PCR methods. The majority of these are present in all PCR or RT-PCR-based techniques. Real-time PCR is susceptible to PCR inhibition by compounds present in certain biological samples. For example, clinical and forensic uses for real-time PCR may be affected by inhibitors found in certain body fluids such as hemoglobin or urea. Food microbiological applications may encounter organic and phenolic inhibitors. To circumvent this problem, alternative DNA polymerases (e.g., *Tfl*, *Pwo*, *Tth*, etc.) that are resistant to particular inhibitors can be used. Other limitations primarily concern real-time PCR-based analysis of gene expression (9–11). Because of the necessary use of RNA in an extra enzymatic step, more problems have the opportunity to occur. RNA itself is extremely labile compared with DNA, and therefore isolation must be carefully performed to ensure both the integrity of the RNA itself and the removal of contaminating nucleases, genomic DNA, and RT or PCR inhibitors. This can be a problem with any sample source, but clinical samples are of special concern because inconsistencies in sample size, collection, storage, and transport can lead to a variable quality of RNA templates. Conversion of RNA to cDNA during the RT reaction is also subject to variability because multiple reverse transcriptase enzymes with different characteristics exist, and different classes of oligonucleotides (e.g., random, poly-dT, or gene-specific primers) can be used to prime RT. Probably the largest present limitation of real-time PCR, however, is not inherent in the technology but rather resides in human error: improper assay development, incorrect data analysis, or unwarranted conclusions [32]. In our experience using real-time PCR for gene expression analysis, real-time PCR primer sets must be designed and validated by stringent criteria to ensure specificity and accuracy of the results. For microbiology, false positives or negatives must be considered when designing an assay to detect pathogens. Amplification and melting curves must be visually inspected while independent calculations based on these curves should be double-checked for accuracy. Real-time PCR gene expression analysis measures mRNA levels and, therefore, only suggests possible changes in protein levels or function rather than demonstrating them. And although there is a tight connection between gene expression and gene product function, this is certainly not always the case, and formal demonstration may be needed for a given research project. Of course, conclusions based on data derived from real-time PCR are best utilized when the biological context is well understood.

Common Applications For Real-Time PCR

Relative and Absolute Quantitation of Gene Expression

To evaluate gene expression, RNA must first be isolated from the samples to be studied. After isolation, RNA is linearly converted to cDNA, which is used for real-time PCR.

Amplification curves are graphed by the software to help determine the "cycle time" at which fluorescence reaches a threshold level (C_T;) This C_T value is inversely proportional to the amount of specific nucleic acid sequence in the original sample. Both relative and absolute quantitation of gene expression utilize the C_T value to quantitate cDNA and thereby determine gene expression. In a perfectly efficient PCR, the amount of amplified product doubles each cycle. Therefore, a difference of 1 between sample C_Ts means that the sample with the lower C_T value had double the target sequence of the other sample; a change in C_T of 2 means a fourfold difference; a change in C_T of 3 means an eightfold difference, and so on ($\Delta C_T = 2^{-\Delta CT}$ fold change).

Relative quantitation measures changes in the steady-state levels of a gene of interest relative to an invariate control gene. Housekeeping genes (e.g., cyclophilin, glyceraldehyde-3-phosphate dehydrogenase, ribosomal protein 36B4, ß-actin, 18S rRNA, transferrin receptor, etc.) that are not expected to change under the experimental conditions serve as a convenient internal standard. Because the absolute quantity of the internal standard is not known, only relative changes can be determined by this method. This may not pose a problem for most research projects because the fold change may be informative irrespective of the absolute value. The limitations of this approach include a lack of absolute quantitation and the necessity for unchanging housekeeping genes as internal standards.

Absolute quantitation attempts a more ambitious task to measure the actual nucleic acid copy number in a given sample. This requires a sample of known quantity (copy number) of the gene of interest that can be diluted to generate a standard curve. This is an external "absolute" standard. Unknown samples are compared with the standard curve for absolute quantitation. The primary limitation to this approach is the necessity of obtaining an independent reliable standard for each gene to be analyzed and then running concurrent standard curves during each assay.

Validation of DNA Microarray Results

Because of the reliability of real-time PCR, many researchers use the above relative or absolute quantitation of gene expression to validate and corroborate the results of printed DNA microarrays or oligonucleotide arrays (e.g., Affymetrix GeneChip). Arrays are used because they allow a researcher to look in an "unbiased" fashion at how experimental manipulation might affect any of the thousands of genes present on the array. Some arrays purport to contain the entire "genome" of a model organism and thus can theoretically be probed to comprehensively determine changes in expression within the entire "transcriptome." The problem is that there can be artifacts, and it is often difficult to get reliable quantitative data or adequate statistical power with present array technology. Thus many researchers choose real-time PCR as a supporting technique to validate and better quantitate the most interesting candidate genes from their arrays [33].

Counting Bacterial, Viral, or Fungal Loads

Real-time PCR can distinguish specific sequences from a complex mixture of DNA. Because of this, it is useful for determining the presence and quantity of pathogen-specific or other unique sequences within a sample.

Identification of Mutations (Or Single Nucleotide Polymorphisms) by Melting Curve Analysis

Real-time PCR is ideally suited for the analysis of mutations, including single nucleotide polymorphisms (SNPs), often replacing other techniques such as sequencing, single-strand conformation polymorphism assays, and restriction fragment-length polymorphism analysis. To detect mutations in the sequence, melting curve analysis is automatically done on the amplified product (amplicon) immediately after PCR thermocycling and fluorescence measurement.

Although any of the above-described fluorescent chemistries will work for detection of mutations, hybridization probes are often used. After PCR is complete, the hybridization probes are attached to the amplicon in tandem, allowing energy to transfer from the donor to acceptor fluorophore, which emits a signal. As the temperature of the reaction vessel increases during the melting curve analysis, the donor probe will dissociate, resulting in a decrease of fluorescence. If there are any mutations in the amplicon (in the hybridization region), the donor probe will bind less strongly and dissociate at a lower temperature. Thus mutations can easily be detected by observing a shift in the melting point of the PCR product. This type of analysis can also be used for genotyping individuals or experimental organisms (i.e., allelic discrimination) [34].

Fields of Real-Time PCR Application

Because of the power of real-time PCR applications, it is already used in many different fields within biomedical research and molecular diagnostics.

Biomedical Research

Real-time PCR has become quite commonplace in basic research within the biomedical sciences. Any time gene expression data is desired for a particular research project, real-time PCR is likely to be used.

Therefore, real-time PCR has impacted a wide variety of topics of study, and the examples of gene expression analyses are innumerable. However, there are additional applications that are particularly useful to basic research. Real-time PCR can be used for genotyping knockout, knockin, and transgenic mouse models or for determining the efficacy of gene knockdown and delivery methods in animals or cell culture systems. With the use of real-time PCR's capability of allelic discrimination, detection of SNPs that may predispose individuals to particular diseases can be determined in populations, thereby facilitating epidemiological studies.

Molecular Diagnostics

Clinical microbiology diagnostic laboratories can use real-time PCR to detect changes in viral load. Because viral load and disease severity are related, real-time PCR can measure the disease progression and efficacy of antiviral therapies. Mutation analyses using melting curves enable individual and epidemiological studies of viral coinfections or quasispecies. Real-time PCR most notably benefits patients when used to detect and identify bacteria. Quick and early detection allows the clinician to immediately prescribe better targeted antibiotic therapies and could, in the long term, help reduce the use of broad-spectrum antibiotics, which may encourage the emergence of antibiotic-resistant strains. Real-time PCR has been used to detect *Mycobacterium tuberculosis*, *Legionella pneumophila*, *Listeria monocytogenes*, and *Neisseria gonorrhoeae*. Mutation (melting curve) analysis has been able to monitor antibiotic resistance among *Staphylococcus aureus*, *Staphylococcus epidermidis*, *Helicobacter pylori*, *Enterococcus faecalis*, and *Enterococcus faecium*. This technology has been extended to quickly detect spores of *Bacillus anthracis*, the well-known causative agent of anthrax and a potential weapon of biological warfare. Real-time PCR is revolutionizing microbiological diagnostics because of the sensitivity of detection and specificity for determination of variants. In addition, there may be substantial time and cost savings over traditional culture methods for determining the quantity of a given pathogen in a clinical specimen. In fact, the Mayo Clinical Microbiology Laboratory has decreased the analytic turnaround time for six different pathogens from a range of 1–14 days by traditional methods to 30–50 min using real-time PCR and has obtained these results with similar or better sensitivities. Of course, there are limitations. Real-time PCR will measure DNA from both live and dead pathogens, whereas traditional culture methods focus on measuring live pathogens. Accordingly, assays for particular pathogens must be responsibly designed and strictly evaluated before conclusions can be made.

The ability to identify specific DNA sequences is also critical in clinical oncology. For example, real-time PCR can be used to detect and sometimes quantify chromosomal translocations or their fusion gene transcripts present in a patient sample for use in the determination of minimum residual disease (MRD) or disease progression. This technique has been used to detect MRD in patients by measuring the AML-1/MTG8 fusion gene product of acute myeloblastic leukemia, several gene product rearrangements of acute lymphoblastic leukemia or the patient response to interferon-α treatment by measuring the BCR-ABL fusion gene product of chronic myeloid leukemia. Real-time PCR can also be used to determine the DNA copy number that can lead to malignancy or be used to analyze gene expression in solid tumors using very small specimens like fine-needle aspirates. Such sensitive and precise detection may prove useful not just for a better understanding of cancer and tumor biology but also for determining more efficacious therapeutic strategies and better stratification of a patient's risk for recurrence of disease.

Real-time PCR can be used to aid drug discovery and characterization by determining the kinetics of target gene expression in response to drugs and the response of transporters or metabolizing enzymes that facilitate distribution or disposal.

Future Applications and Perspectives

Any need for fast and precise measurement of small amounts of nucleic acids represents a potential future niche for real-time PCR-based innovations.

As machines become faster, cheaper, smaller, and easier to use through competition, standardized assay development, and advances in microfluidics, optics, and thermocycling, more in-field application needs are likely to be filled. In the commercial food industry and agriculture, real-time PCR will likely see expanded use for the detection and identification of microbes, parasites, or genetically modified organisms. Forensics will benefit from real-time PCR's sensitivity, specificity, and speed, especially because time is crucial to many criminal investigations and specimen size may be limited. Reduced cost and increased portability open the door for the diagnosis of diseases in remote areas along with on-site epidemiological studies and may facilitate the transfer of needed scientific technologies to developing countries, thereby contributing to their "scientific capacity". Because the demand to measure gene expression is unlikely to wane as long as biomedical science is thriving, new generations of real-time PCR machines are likely to be developed. Much like computers, earlier generations of machines should be relatively inexpensive and therefore increase global access to the technology [35,36,37]. Hopefully, this technology will enter the classroom to enable life science educators to better equip students and encourage them to consider careers in science. Teaching real-time PCR could be used both as a platform to introduce key concepts in molecular biology as well as a chance to give students confidence by successfully learning and implementing scientifically relevant skills. This would be a great way to increase hands-on learning, which may be a key component to improving biology education in the United States.

CONCLUSION

Real-time PCR generates a focused look at the "transcriptome," enabling researchers to better understand the transcriptional programs that underlie physiology, pathophysiology, and development. Understanding gene expression narrows the gap in our knowledge between the "instructions" (the genome) and the "functions" (gene products) of biology.

The evolution of the science and technology of molecular biology can be viewed chronologically as epochs hallmarked by the maturation of study of particular biomolecules. Extensive, and in some ways comprehensive, analysis of DNA (genomics) was first realized in the Human Genome Project. Now analysis of RNA expression profiles (transcriptomics) is reaching maturity. The imminent future promises great leaps forward in the analysis of proteins (proteomics), whose technology is the most rapidly growing, and in the analysis of biological lipids or metabolic intermediates (lipomics or metabolomics, respectively).

Hopefully, the pieces of the biomolecular puzzle can be put together, leading to a more holistic understanding of biology. To form a coherent picture, however, parallel advances in data acquisition, compilation, and analysis will be necessary to help deal with the enormity of data. The promise of such extensive knowledge of biological systems is staggering but will certainly require dedicated individuals in all biomedical fields to figure out how best to utilize the new technologies and the information produced by them.

References

[1] Ballard AL, Fry NK, Chan L, Surman SB, Lee JV, Harrison TG, and Towner KJ. (2000) Detection of *Legionella pneumophila* using a real-time PCR hybridization assay. *J Clin Microbiol* 38: 4215–4218.

[2] Barthe C, Mahon FX, Gharbi MJ, Faberes C, Bilhou-Nabera C, Hochhaus A, Reiffers J, and Marit G. (2001) Expression of interferon-α (IFN-α) receptor 2c at diagnosis is associated with cytogenetic response in IFN-α-treated chronic myeloid leukemia. *Blood* 97: 3568–3573.

[3] Brand CL, Sturis J, Gotfredsen CF, Fleckner J, Fledelius C, Hansen BF, Andersen B, Ye JM, Sauerberg P, and Wassermann K. (2003) Dual PPARα/γ activation provides enhanced improvement of insulin sensitivity and glycemic control in ZDF rats. *Am J Physiol Endocrinol Metab* 284: E841–E854.

[4] Brazeau DA. (2004) Combining genome-wide and targeted gene expression profiling in drug discovery: microarrays and real-time PCR. *Drug Disc Today* 9: 838–845.

[5] Brown PJ, Stuart LW, Hurley KP, Lewis MC, Winegar DA, Wilson JG, Wilkison WO, Ittoop OR, and Willson TM. (2001) Identification of a subtype selective human PPARα agonist through parallel-array synthesis. *Bioorg Med Chem* 11: 1225–1227.

[6] Brown PO and Botstein D. (1999) Exploring the new world of the genome with DNA microarrays. *Nat Genet* 21: 33–37.

[7] Bustin SA. (2000) Absolute quantification of mRNA using real-time reverse transcription polymerase chain reaction assays. *J Mol Endocrinol* 25: 169–193.

[8] Bustin SA. (2002) Quantitation of mRNA using real-time reverse transcription PCR (RT-PCR): trends and problems. *J Mol Endocrinol* 29: 23–39.

[9] Bustin SA and Nolan T. (2004) Pitfalls of quantitative real-time reverse-trasncription polymerase chain reaction. *J Biomol Tech* 15: 155–166.

[10] Cleary TJ, Roudel G, Casillas O, and Miller N. (2003) Rapid and specific detection of *Mycobacterium tuberculosis* by using the Smart Cycler instrument and a specific fluorogenic probe. *J Clin Microbiol* 41: 4783–4786.

[11] Dheda K, Huggett JF, Bustin SA, Johnson MA, Rook G, and Zumla A. (2004) Validation of housekeeping genes for normalizing RNA expression in real-time PCR. *Biotechniques* 37: 112–119.

[12] Edwards K, Logan J, and Saunders N. (2004) *Real-time PCR: an Essential Guide.* Norfolk, UK: Horizon Bioscience.

[13] Ferre P. (2004) The biology of peroxisome proliferator-activated receptors: relationship with lipid metabolism and insulin sensitivity. *Diabetes* 53: S43–S50.

[14] Gibson JR, Saunders NA, Burke B, and Owen RJ. (1999) Novel method for rapid determination of clarithromycin sensitivity in Helicobacter pylori. *J Clin Microbiol* 37: 3746–3748.

[15] Ginzinger DG, Godfrey TE, Nigro J, Moore DH, Suzuki S, Pallavicini MG, Gray JW, and Jensen RH. (2000) Measurement of DNA copy number at microsatellite loci using quantitative PCR analysis. *Cancer Res* 60: 5405–5409.

[16] Harris E. (2004) Building scientific capacity in developing countries. *EMBO J* 5: 7–11.

[17] Hein I, Klein D, Lehner A, Bubert A, Brandl E, and Wagner M. (2001) Detection and quantification of the iap gene of *Listeria monocytogenes* and *Listeria innocua* by a new real-time quantitative PCR assay. *Res Microbiol* 152: 37–46.

[18] Higuchi R, Dollinger G, Walsh PS, and Griffith R. (1992) Simultaneous amplification and detection of specific DNA sequences. *Biotechnology* 10: 413–417.

[19] Higuchi R, Fockler C, Dollinger G, and Watson R. (1993) Kinetic PCR analysis: real-time monitoring of DNA amplification reactions. *Biotechnology* 11: 1026–1030.

[20] Klein D. (2002) Quantification using real-time PCR technology: applications and limitations. *Trends Mol Med* 8: 257–260.

[21] Krauter J, Heil G, and Ganser A. (2001) The AML1/MTG8 fusion transcript in t(8;21) positive AML and its implication for the detection of minimal residual disease; malignancy. *Hematology* 5: 369–381.

[22] Kuhne BS and Oschmann P. (2002) Quantitative real-time PCR using hybridization probes and imported standard curves for cytokine gene expression analysis. *Biotechniques* 33: 1078–1084.

[23] Kurrasch DM, Huang J, Wilkie TM, and Repa JJ. (2004) Quantitative real-time PCR measurement of regulators of G-protein signaling (RGS) mRNA levels in mouse tissues. *Methods Enzymol* 389: 3–15.

[24] Le Pecq JB and Paoletti C (1996). A new fluorometric method for RNA and DNA determination. *Anal Biochem* 17: 100–107.

[25] Lunge VR, Miller BJ, Livak KJ, and Batt CA. (2002) Factors affecting the performance of 5' nuclease PCR assays for *Listeria monocytogenes* detection. *J Microbiol Methods* 51: 361–368.

[26] Luu-The V, Paquet N, Calvo E, and Cumps J. (2005) Improved real-time RT-PCR method for high-throughput measurements using second derivative calculation and double correction. *Biotechniques* 38: 287–293.

[27] Mackay IM. (2004) Real-time PCR in the microbiology laboratory. *Clin Microbiol Infect* 10: 190–212.

[28] Mackay IM, Arden KE, and Nitsche A. (2002) Real-time PCR in virology. *Nucleic Acids Res* 30: 1292–1305.

[29] Makino SI, Cheun HI, Watari M, Uchida I, and Takeshi K. (2001) Detection of anthrax spores from the air by real-time PCR. *Lett Appl Microbiol* 33: 237–240.

[30] Martineau F, Picard FJ, Lansac N, Menard C, Roy PH, Ouellette M, and Bergeron MG. (2000) Correlation between the resistance genotype determined by multiplex PCR assays and the antibiotic susceptibility patterns of *Staphylococcus aureus* and *Staphylococcus epidermidis*. *Antimicrob Agents Chemother* 44: 231–238.

[31] Mitchell P. (2001) Microfluidics–downsizing large-scale biology. *Nat Biotechnol* 19: 717–721.

[32] Mullis KB. (1990) The unusual origin of the polymerase chain reaction. *Sci Am* 262: 56–61.

[33] Ohnmacht GA, Wang E, Mocellin S, Abati A, Filie A, Fetsch P, Riker AI, Kammula US, Rosenberg SA, and Marincola FM. (2001) Short-term kinetics of tumor antigen expression in response to vaccination. *J Immunol* 167: 1809–1820.

[34] Olson M, Hood L, Cantor C, and Botstein D. (1989) A common language for physical mapping of the human genome. *Science* 245: 1434–1435.

[35] Pongers-Willemse MJ, Verhagen OJ, Tibbe GJ, Wijkhuijs AJ, de Haas V, Roovers E, van der Schoot CE, and van Dongen JJ. (1998) Real-time quantitative PCR for the detection of minimal residual disease in acute lymphoblastic leukemia using junctional region specific TaqMan probes. *Leukemia* 12: 2006–2014.

[36] Powledge TM. (2004) The polymerase chain reaction. *Adv Physiol Educ* 28: 44–50.

[37] Reischl U, Wittwer CT, and Cockerill F. (2002) *Rapid Cycle Real-time PCR: Methods and Applications; Microbiology and Food Analysis.* New York: Springer-Verlag.

In: Advances in Biotechnology ISBN: 978-1-62417-842-9
Editors: H. K. Dhingra, P. N. Jha et al.

Chapter 8

RNA INTERFERENCE

Sushma Manral* and Teresa Biswas
Department of Biochemistry, Maulana Azad Medical College, Delhi, India

ABSTRACT

RNA interference (RNAi) has become a powerful tool not only in the field of functional genomics but also in disease therapeutics, in particular cancer. It is mediated by RNA duplexes which causes sequence specific gene silencing. Double stranded RNA (short hairpin RNA, shRNA) upon cleaved by Dicer, generates small interfering RNA (siRNA). The siRNA interacts with RISC (RNA induced silencing complex) and inhibit the action of target mRNA by either its degradation or its repression or chromatin modifications. Micro RNA (miRNA) also has similar function but is processed from primary transcription unit and involves one more enzyme, Drosha. Many clinical trials are under process for the use of RNAi in various diseases like cancer, ocular diseases, metabolic diseases and many infectious diseases. As RNAi transcription cassettes needs to be delivered to cells, various virus based vectors are under study for their use as delivery vehicles for the same.

Keywords: RNAi, siRNA, miRNA

* Corresponding Author: Sushma Manral. Email: sushmamanral@gmail.com.

Introduction

RNA interference (RNAi) is an extremely powerful tool for regulation of genetic expression mediated by short RNA duplexes that silences genes in sequence specific manner. RNA interference involves either the specific degradation of mRNA or repression of translation of the mRNA containing sequences complimentary to those present in double stranded RNA [1-3]. It may also act by silencing the promoter that directs the expression of mRNA through chromatin modification [4].

Although discovered serendipitously, the technique is now extensively exploited as a research tool and for searching treatment modalities for a number of diseases.

Micro RNA (miRNA) as well as small interfering RNA (siRNA) act through the pathway of RNA interference. These two differ from each other in terms of their origin. siRNA are produced from dsRNA precursors (short hairpin RNA, shRNA) in vivo or are introduced artificially. But miRNA are processed from precursor mRNA that is encoded by genes expressed in cells where the miRNA serve a regulatory function. Both small RNA are derived from longer mRNA but the processing of miRNA involves two enzymes Drosha and Dicer but siRNA involves only Dicer [5-7].

Synthesis of miRNA

miRNA are basically encoded by genome as longer transcript and possess characteristic structure which helps to identify them and predict the target gene they might regulate. The functional miRNA, which is 21-22 nucleotides long, is generated from primary mRNA transcript (pri-miRNA) by the action of Drosha, a catalytic enzyme. The pri-miRNA carries a hairpin shaped secondary structure.

The cleavage by Drosha liberates stem loop called pre-miRNA. The pre-miRNA may be encoded in the coding region, leader sequence or introns.

The enzyme Dosha is a member of RNaseIII family of enzymes. It makes two cuts on primary transcript and liberates the stem loop region of pre-miRNA. Drosha acts in conjunction with another protein (called Pasha in some organism and DGCR8 in others) and forms microprocessor complex. The pre-miRNA generated by Drosha is 65-70 nucleotides long and contains 2 nucleotides overhang on the 3 ends of the ds RNA product. As Drosha resides in the cell's nucleus, the generation of pre-miRNA from pri-miRNA also occurs in nucleus.

The pre-miRNA is then exported out of nucleus by exportin-5/RanGTP into the cytoplasm where it is acted upon by another cleavage enzyme, Dicer [5-10]. Dicer has two RNase III domains and a ds RNA- binding domain called the PAZ domain (named for three proteins, Piwi, Argonaute, and Zwille, which contain this domain). Dicer acts on any ds RNA, regardless of the sequence, and cleave the ds RNA 22 nucleotides from its end. The PAZ domain anchors the ds RNA by its 3 overhang and also places the substrate RNA in a fashion so as to be cleaved 22 nt away from its end.

PRODUCTION OF siRNA

The precursor of siRNA is a long dsRNA (shRNA) in contrast to the hairpin precursor of miRNA. As a consequence of this, Drosha is not needed for the generation of siRNA. But dicer is still needed for the generation of 21 to 22 nucleotide long siRNA.

MECHANISM OF RNA INTERFERENCE

MiRNA and siRNA both acts through same machinery which includes a complex called the RNA-induced silencing complex (RISC). The RISC, in addition to siRNA or miRNA, contains various proteins including a member of Argonaute family. Exogenous synthetic siRNA or endogenously expressed siRNA are directly incorporated into RISC, bypassing the action of Dicer. A helicase in RISC unwinds the duplex siRNA or miRNA to give a *guide RNA*-the strand that provides specificity to the RISC, and a *passenger RNA*-the strand which gets degraded (Figure 1).

The mature RISC (RISC + guide RNA) is then directed to the target RNAs containing the sequence complimentary to the guide RNA. The target RNA is then either degraded or its translation is inhibited, depending upon how closely the guide RNA matches with the target RNA. If the match is perfect, the target RNA is degraded, but if the match is imperfect then translation is inhibited. Usually miRNA induces repression of transcription by imperfect pairing and siRNA causes degradation of target mRNA by perfect pairing with the target mRNA.

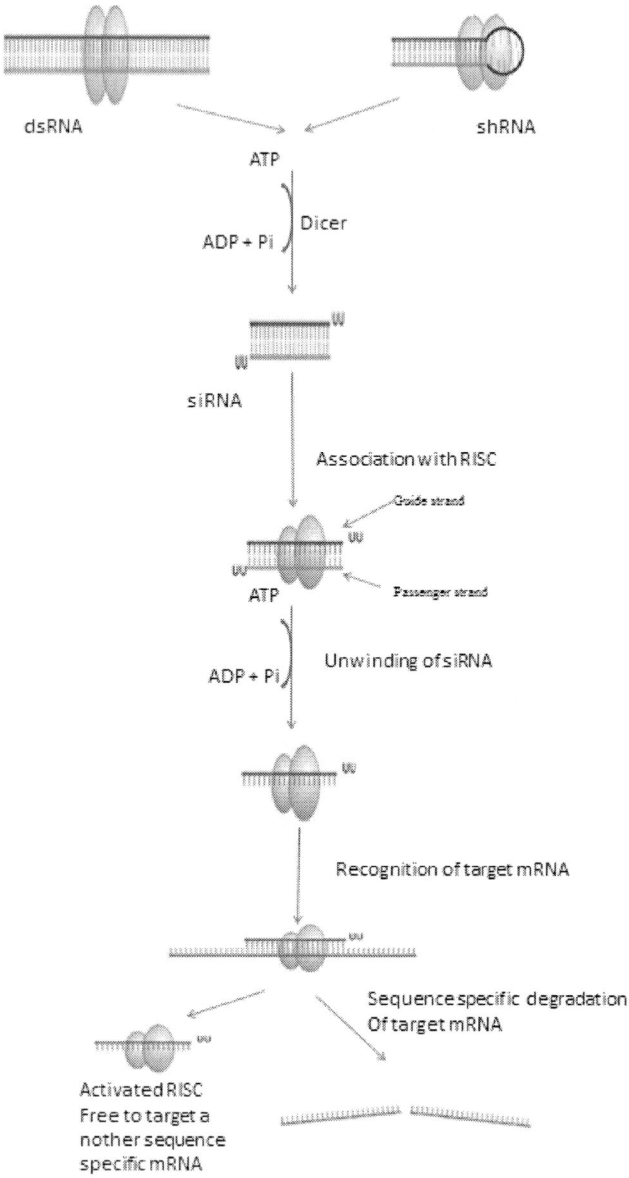

Figure 1. Working mechanism of siRNA.

The repression of translation is still under scrutiny, but it seems that miRNA leads to sequestration of target RNA in so called processing bodies (P-bodies) in the cytoplasm where translation is repressed. miRNA also causes disruption of poly A tail of target mRNA resulting in disruption of translation. In cases where the target mRNA is degraded, Argonaute causes the initial cleavage and is hence called the "slicer" and the process of RNA degradation as "slicing". A RISC when directed into the nucleus recruits other proteins and leads to chromatin modifications around the promoter of the gene complimentary to guide RNA, thereby silencing the gene [4, 9].

RNA interference is extremely efficient. Very small amount of dsRNA are enough for inducing a complete shutdown of target gene expression. RNA dependent RNA polymerase (RdRP) plays central role in providing efficiency to RNA interference. The RdRP converts single stranded RNAs into dsRNAs. The RdRP can synthesise dsRNA in primer independent and dependent manner. The latter involves use of small RNAs (miRNA or siRNA) as the primer and is actually required for the amplification and formation of secondary siRNAs.

As the resulting secondary siRNAs are in antisense orientation, these can direct the negative regulation of target transcripts from the same gene or complimentary sequences thus, amplifying the gene silencing [9-11].

Application of RNA Interference

RNA interference and cancer:

Cancer is characterised by deregulation of cell proliferation and apoptosis due to epigenetic changes or mutations [12]. There are two classes of genes: tumour suppressor genes and oncogenes, loss of function or gain of function mutations of which, respectively, are tumourogenic. RNAi is being explored as a potential anticancer therapy by inhibiting the expression of genes involved in oncogenesis [13]. Theoretically, iRNA machinery can be exploited for silencing any part of genome by use of appropriately designed siRNA. Ovarian cancer and bone cancer are currently investigated and successfully treated with siRNA in vivo [14, 15]. Experimentally, the effective targets for iRNA use in cancer are:

1. Genes involved in cancer assosciated cellular pathways, viz. Oncogenes (particularly fusion oncogene) or antiapoptotic genes or tumour suppressor genes. For example; BCR-ABL fusion gene formed by translocation of Philedelphia chromosome, is involved in leukaemic transformation in chronic myelogenous leukemia and Ph-positive acute lymphoblastic leukemia [16]. The siRNA directed

against BCR-ABL transcript have displayed silencing of the fusion transcript without affecting the normal cABL and cBCR transcripts expression [13, 17]. Similar studies in pancreatic and colon carcinoma has revealed that RNAi specific for K-RAS$^{1/2}$ oncogene reduces the K-RAS$^{1/2}$ transcript levels and resulting tumourogenicity [18, 19].

2. Genes that play a role in tumour host interaction such as factors involved in innate immunity or angiogenesis.
3. Genes involved in imparting resistance to chemotherapy or radiotherapy [20, 21].
4. The role of iRNA in miRNA caused malignancies is also very interesting therapy field. miRNA mediated effects by regulating the human gene transcripts have been implicated in number of human diseases. Manipulation of miRNA levels by the use of gene therapy provides an attractive approach for controlling gene expression [22-24].

RNA Interference and Infectious Diseases

HIV is the first infectious agent to be studied for the iRNA therapy as its life cycle and expression pattern is well understood [4]. Although iRNA proved to be successful in inhibiting viral encoded RNAs in cell culture, high rate of viral mutation limits its clinical application [26]. Therefore, in order to overcome this limitation, down regulation of cellular cofactors like NF-κB, HIV receptor CD4 and the co-receptors CXCR4 and CCR5, required for HIV infection proved as an attractive alternative for iRNA [27-29]. Examples of successful iRNA therapy for combating viral pathogenesis are: Hepatitis B virus (HBV) [30, 31], Human Papilloma Virus [32], Severe Acute Respiratory Syndrome (SARS) Coronavirus [33], and Respiratory Syncytial Virus (RSV) infections [34].

RNA Interference and Ocular Diseases

The eye is an isolated chamber and easily approachable and thus ocular diseases prove to be an ideal target for siRNA based therapies [35]. The amount of siRNA needed to be injected into vitreous cavity is very small and gets readily diffused throughout the eye [36]. Age related macular degeneration (AMD) was the first target for siRNA clinical trials. VEGF (vascular endothelial growth factors) is believed to play primary role in vascular overgrowth seen in wet type of AMD.

The clinical trial involved intravitreal injection of of siRNA targeting the VEGF. In 2004, the company Sirna presented the first ever clinical data for an iRNA based drug- the compound AGN-745 directed against VEGF receptor 1, for the treatment of AMD [37, 38]. According to clinical investigators, all patients enrolled for the trial have experienced visual acuity stabilisation during the trial [39].

RNA Interference and Other Therapeutically Relevant Fields

Metabolic diseases, cardiac diseases, human neurodegenerative diseases and inherited genetic diseases are the other relevant fields for iRNA therapy. PCSK9 is a member of serine protein convertase family and has been implicated in hypercholesteronemia. A siRNA targeted against PCSK9 lowered plasma cholesterol thus offering a good approach to control this disease [40]. Many clinical trials for coronary artery diseases using siRNA against c-myc responsible for closure of arteries after stent insertion are also underway [41]. Role of siRNA in controlling liver fibrosis, although in rats, have highlighted a new therapeutic approach for reversing human liver cirrhosis [42].

Delivery of Interfering RNA

Gene silencing by RNAi can be achieved by two ways:

1. Cytosolic delivery of chemically synthesised 21nt siRNA duplexes with 2nt 3' overhangs: This mode of administration is transient, as the intracellular concentration of siRNA gets diluted with cell division. Also duplex siRNAs are negatively charged polymers and therefore do not penetrate hydrophobic cellular membranes without assisting carriers. Moreover, RNases rapidly degrades the unprotected and unmodified siRNAs.

2. Nuclear delivery of shRNA: here, the shRNA is processed like endogenous interfering RNA resulting in stable gene downregulation. For chronic diseases, this long term gene downregulation is required.

Thus, an ideal delivery vehicle should facilitate endosomal or lysosomal escape and, in case of shRNA, delivery through the nuclear membrane. Therefore viruses and vectors derived from them has become a major delivery system for shRNA.

RNAi can be induced in cells by intracellular expression of miRNAs either by shRNAs or synthetic miRNAs [25, 43, 44]. shRNA constitute a sense and antisense sequences connected by a loop of unpaired nucleotides while the stem loop of miRNA are transcribed from pri-mi RNA [45].

Artificial miRNA resemble more to the endogenous RNAi substrates and are more amenable to Pol-II transcription and thus promises a definite route for gene therapies [46, 47].

Viral Delivery of shRNA Expression Cassettes

The combination of viral gene therapy and RNAi proves to be a powerful approach in places where spatiotemporal control over gene or persistence silencing of target gene is very important. The action of viral RNAi vectors is influenced by the fact that promoter directed shRNAs induces RNAi machinery [47]. This involves cloning of an oligonucleotide containing siRNA sequence into plasmid or viral vectors which got endogenously expressed to shRNA. The latter subsequently got processed in cytoplasm to siRNA. As shRNA expression cassesttes are extremely small in size, it is possible to pack them in smallest viral vectors.

The various viral vector systems are:

1. Adenovirus vectors: Adenoviruses are medium sized non enveloped viruses with a nucleocapsid and a linear dsDNA genome. They replicate inside the nucleus of human cells without getting incorporated into the host's genome. They can carry 8-30kb of foreign DNA. Features of adenovirus which attracts its use as a vector are: infection of both dividing and non-dividing cells, high levels of transgene expression, and ability to grow to high titres in vitro. As adenovirus is a causative agent for respiratory infections in children,

the seropositivity for the same is quite common posing a drawback for adenovirus therapy.

Moreover, Adenovirus shows no specific tissue tropism and may induce strong immune reaction on repeated administration when used as a vector. Many studies have revealed adenoviral vectors as a common mode of delivery of shRNA expression cassettes into target cells in vitro [43, 48].

2. Adeno-associated viruses (AAV): It is one of the smallest viruses with single stranded DNA genome and is non pathogenic to humans. Although the packaging capacity is 5kb but is sufficient to accommodate atleast eight individual shRNA expression cassettes [49]. The virus is replication defective and needs adenovirus for replication. But recombinant AAV with single adenovirus genes are replicative efficient and proves better vector than adenovirus as their pseudotyping permits tissue specific delivery of shRNA expression cassettes [50, 51]. Moreover, AAV vectors possess low reactivity to cellular immune response.

   As both adenovirus and AAV vectors are non-integrative so are free from insertional mutagenesis. But this property also limits its long term effect as genetic information is lost with repetitive cell division.
3. Retrovirus vectors: Retroviruses are the single stranded RNA viruses which replicate through double standed DNA intermediates. They integrate their genome stably into host cell's DNA thereby allowing long term expression of inserted therapeutic genes.viral genome codes for atleast three genes: gag (codes cor proteins), pol (codes for reverse transcriptase) and env (codes for envelop proteins). As retroviral entry and gene integration does not require viral proteins, so genes encoding for viral proteins can thus be replaced by foreign sequences [52]. By changing the env gene or its products, the target tissue range can be manipulated [53-56]. Retroviruses were among the first vectors used for delivery of shRNA expressing plasmids. Many studies have used MLV (murine leukemic virus), a retrovirus, as transfer vehicles for shRNA expression cassettes [57, 58].
4. Lentivirus vectors: Lentiviruses (LV) are a subclass of retroviruses which carry two copies of single stranded RNA genome in an enveloped capsid. These are capable of transducing both dividing and non-dividing cells. Lentivirus vectors accommodate large amount of DNA and evoke host immune reaction minimally [59]. These are also one of the most commonly used vehicles for shRNA delivery and also

have been used for generation of transgenic animals [60]. Many companies offer this method for in vitro experiments.

5. Baculoviruses: These are insect viruses incapable of replicating in humans. Baculovirus based delivery methods for shRNA is being used for taetting viral infections, for example; HCV [61], influenza virus A and B [62, 63].

CONCLUSION

RNAi is becoming a powerful technology which can be exploited for gene therapy in days to come. But the assurance of successful results not only depends upon the technology but also on other factors like target specific delivery of shRNA expression cassettes. The use of viral vectors looks promising but their effectiveness is still under study.

REFERENCES

[1] Hannon G.J. RNAi: A guide to gene silencing. Cold Spring Harbour Laboratory Press, Cold Spring Harbour, New York. 2003,

[2] Baulcombe D. RNA silencing. Trends Biochem. Sci. 2005,30: 290-293.

[3] Cook A. And Conti E. Dicer measures up. *Nat. Struct. Mol. Biol.,* 2006, 13:190-192.

[4] Hannon G. J., Rossi J. Unlocking the potential of the human genome with RNA interference. *Nature,* 2004,431:371–378.

[5] Lund E., Guttinger S., Calado A., Dahlberg J. E. and Kutay U. Nuclear export of microRNA precursors. *Science,* 2004,303, 95–98.

[6] Pei Y. And Yuschl T. On the art of identifying effective and specific siRNAs. *Nat. Methods,* 2006, 3: 670-676.

[7] Peters L. And Meister G. Argonaute protens: Mediators of RNA silencing. *Mol. Cell,* 2007,26: 611-623.

[8] Rana T.M. Illuminating the silence: understanding the structure and function of small RNAs. *Nat. Rev. Mol. Cell Biol.,* 2007,8:23-36

[9] Tang G. siRNA and miRNA: An insight into RISCs. *Trends Biochem. Sci.,* 2005,30: 106-114.

[10] Tolia N.H. and Joshua-Tor L. Slicer and the Argonautes. *Nat. Chem. Biol.,* 2007,3: 36-43.

[11] Valencia-Sanchez M.A., Liu J., Hannon G.J., Parker R. Control of translation and mRNA degradation by miRNAs and siRNAs. *Genes Dev.,* 2006,20: 515-524.

[12] Weiler J., Hunziker J. and Hall J. Anti-miRNA oligonucleotides (AMOs): ammunition to target miRNAs implicated in human disease? *Gene Ther.*, 2006,13, 496–502.

[13] Scherr M., Battmer K., Winkler T., Heidenreich O., Ganser A. And Eder M. Specific inhibition of *bcr-abl* gene expression by small interfering RNA. *Blood,* 2003,101, 1566–1569.

[14] Halder J, Kamat AA, Landen CN Jr, Han LY, Lutgendorf SK, Lin YG, et al: Focal adhesion kinase targeting using in vivo short interfering RNA delivery in neutral liposomes for ovarian carcinoma therapy. *Clin. Cancer Res.,* 2006, 12:4916-4924.

[15] Takeshita F, Minakuchi Y, Nagahara S, Honma K, Sasaki H, Hirai K, et al: Efficient delivery of small interfering RNA to bone-metastatic tumors by using atelocollagen in vivo. *Proc. Natl. Acad. Sci. USA,* 2005, 102:12177-12182.

[16] Stevenson M. 2004. Therapeutic potential of RNA interference. *NEJM,* 351, 1772–1777.

[17] Wohlbold L., van der Kuip H., Miething C., Vornlocher H. P., Knabbe C., Duyster J. and Aulitzky W. E. inhibition of bcr-abl gene expression by small interfering RNA sensitizes for imatinib- mesylate (STI571). *Blood,* 2003,102:2236–2239.

[18] Brummelkamp T. R., Bernards R. and Agami R. Stable suppression of tumorigenicity by virus-mediated RNA interference. *Cancer Cell,* 2002,2: 243–247.

[19] Wilda M., Fuchs U., Wossmann W. and Borkhardt A. Killing of leukemic cells with a BCR/ABL fusion gene by RNA interference (RNAi). *Oncogene,* 2002,21: 5716–5724.

[20] Hanahan D, Weinberg RA: The hallmarks of cancer. *Cell,* 2000, 100:57-70.

[21] Ellis LM, Hicklin DJ: Resistance to Targeted Therapies: Refining Anticancer Therapy in the Era of Molecular Oncology. *Clin. Cancer Res.,* 2009, 15:7471-7478.

[22] Marquez RT, McCaffrey AP: Advances in microRNAs: implications for gene therapists. *Hum. Gene Ther.,* 2008, 19:27-38.

[23] Calin GA, Croce CM: MicroRNA signatures in human cancers. *Nat. Rev. Cancer,* 2006, 6:857-866.

[24] Calin GA, Croce CM: MicroRNA-cancer connection: the beginning of a new tale. *Cancer Res.,* 2006, 66:7390-7394.

[25] Elbashir SM, Harborth J, Lendeckel W, Yalcin A, Weber K, Tuschl T: Duplexes of 21-nucleotide RNAs mediate RNA interference in cultured mammalian cells. *Nature,* 2001, 411:494-8.

[26] Boden D., Pusch O., Lee F., Tucker L. and Ramratnam B. Human immunodeficiency virus type-1 escape from RNA interference. *J. Virol.*, 2003,77:11531–11535.

[27] Surabhi R. M., Gaynor R. B. RNA interference directed against viral and cellular targets inhibits human immunodeficiency virus type-1 replication. *J. Virol.*, 2002,76, 12963–12973.

[28] Novina C. D., Murray M. F., Dykxhoorn D. M., Beresford P. J., Riess J., Lee S. K. et al. siRNA-directed inhibition of HIV-1 infection. *Nature Med.*, 2002,8:681–686.

[29] Martinez M. A., Guti´errez A., Armand-Ug´on M., Blanco J., Parera M., G´omez J. *et al.* Suppression of chemokine receptor expression by RNA interference allows for inhibition of HIV-1 replication. *AIDS,* 2002b,16:2385–2390.

[30] Morrissey DV, Lockridge JA, Shaw L, Blanchard K, Jensen K, Breen W, et al: Potent and persistent in vivo anti-HBV activity of chemically modified siRNAs. *Nat. Biotechnol.,* 2005, 23:1002-1007.

[31] Song E, Lee SK, Wang J, Ince N, Ouyang N, Min J, et al: RNA interference targeting Fas protects mice from fulminant hepatitis. *Nat. Med.,* 2003, 9:347-351.

[32] Niu XY, Peng ZL, Duan WQ, Wang H, Wang P: Inhibition of HPV 16 E6 oncogene expression by RNA interference in vitro and in vivo. *Int. J. Gynecol. Cancer,* 2006, 16:743-751.

[33] Li BJ, Tang Q, Cheng D, Qin C, Xie FY, Wei Q, et al: Using siRNA in prophylactic and therapeutic regimens against SARS coronavirus in Rhesus macaque. *Nat. Med.,* 2005, 11:944-951.

[34] Bitko V, Musiyenko A, Shulyayeva O, Barik S: Inhibition of respiratory viruses by nasally administered siRNA. *Nat. Med.,* 2005, 11:50-55.

[35] Wallach T. Acuity pharmaceuticals announces completion of financing round. *Acuity Pharmaceuticals,* 2004, 215:966–6181.

[36] Campochiaro P. A. Potential applications for RNAi to probe pathogenesis and develop new treatments for ocular disorders. *Gene Ther.*, 2006,13:559–562.

[37] McFarland TJ, Zhang Y, Appukuttan B, Stout JT: Gene therapy for proliferative ocular diseases. Expert Opin Biol Ther 2004, 4:1053-1058.

[38] Check E: A crucial test. *Nat. Med.,* 2005, 11:243-244.
[39] Cejka D., Losert D.,Wacheck V. Short interfering RNA (siRNA): tool or therapeutic?. *Clin. Sci.*, 2006,110:47–58.
[40] Abifadel M, Varret M, Rabes JP, Allard D, Ouguerram K, Devillers M, et al: Mutations in PCSK9 cause autosomal dominant hypercholesterolemia. *Nat. Genet.,* 2003, 34:154-156.
[41] Li FY, Cheng NS, Cheng JQ, Mao H, Jiang LS, Li N, et al: Treatment of chronic proliferative cholangitis with c-myc shRNA. *World J. Gastroenterol.,* 2009, 15:95-101.
[42] Sato Y, Murase K, Kato J, Kobune M, Sato T, Kawano Y, et al: Resolution of liver cirrhosis using vitamin A-coupled liposomes to deliver siRNA against a collagen-specific chaperone. *Nat. Biotechnol.,* 2008, 26:431-442.
[43] Xia H, Mao Q, Paulson HL, Davidson BL: siRNA-mediated gene silencing in vitro and in vivo. *Nat. Biotechnol.,* 2002, 20:1006-1010.
[44] Zeng Y, Wagner EJ, Cullen BR: Both natural and designed micro RNAs can inhibit the expression of cognate mRNAs when expressed in human cells. *Mol. Cell,* 2002, 9:1327-1333.
[45] Cai X, Hagedorn CH, Cullen BR: Human microRNAs are processed from capped, polyadenylated transcripts that can also function as mRNAs. *RNA,* 2004, 10:1957-1966.
[46] Chung KH, Hart CC, Al Bassam S, Avery A, Taylor J, Patel PD, et al: Polycistronic RNA polymerase II expression vectors for RNA interference based on BIC/miR-155. *Nucleic Acids Res.,* 2006, 34:e53.
[47] Brummelkamp TR, Bernards R, Agami R: A System for Stable Expression of Short Interfering RNAs in Mammalian Cells. *Science,* 2002, 296:550-3.
[48] Relph KL, Harrington KJ, Pandha H: Adenoviral strategies for the gene therapy of cancer. *Semin. Oncol.,* 2005, 32:573-582.
[49] Grimm D, Pandey K, Kay MA: Adeno-associated virus vectors for short hairpin RNA expression. *Methods Enzymol.,* 2005, 392:381-405.
[50] Matsushita T, Elliger S, Elliger C, Podsakoff G, Villarreal L, Kurtzman GJ, et al: Adeno-associated virus vectors can be efficiently produced without helper virus. *Gene Ther.,* 1998, 5:938-945.
[51] Haynes C, Erlwein O, Schnierle BS: Modified envelope glycoproteins to retarget retroviral vectors. *Curr. Gene Ther.,* 2003, 3:405-10.
[52] Mann R, Mulligan RC, Baltimore D: Construction of a retrovirus packaging mutant and its use to produce helper-free defective retrovirus. *Cell,* 1983, 33:153-159.

[53] Haynes C, Erlwein O, Schnierle BS: Modified envelope glycoproteins to retarget retroviral vectors. *Curr. Gene Ther.,* 2003, 3:405-10.

[54] Erlwein O, Buchholz CJ, Schnierle BS: The proline-rich region of the ecotropic Moloney murine leukaemia virus envelope protein tolerates the insertion of the green fluorescent protein and allows the generation of replication-competent virus. *J. Gen. Virol.,* 2003, 84:369-73.

[55] Sliva K, Erlwein O, Bittner A, Schnierle BS: Murine leukemia virus (MLV) replication monitored with fluorescent proteins. *Virol. J.,* 2004, 1:14.

[56] Sliva K, Schnierle BS: Targeting retroviral vector entry by host range extension. *Gen. Ther. Mol. Biol.,* 2004, 8:335-342.

[57] Paddison PJ, Silva JM, Conklin DS, Schlabach M, Li M, Aruleba S, et al: A resource for large-scale RNA-interference-based screens in mammals. *Nature,* 2004, 428:427-431.

[58] Berns K, Hijmans EM, Mullenders J, Brummelkamp TR, Velds A, Heimerikx M, et al: A large-scale RNAi screen in human cells identifies new components of the p53 pathway. *Nature,* 2004, 428:431-437.

[59] al Yacoub N, Romanowska M, Haritonova N, Foerster J: Optimized production and concentration of lentiviral vectors containing large inserts. *J. Gene Med.,* 2007, 9:579-584.

[60] Naldini L, Blomer U, Gage FH, Trono D, Verma IM: Efficient transfer, integration, and sustained long-term expression of the transgene in adult rat brains injected with a lentiviral vector. *Proc. Natl. Acad. Sci. USA,* 1996, 93:11382-11388.

[61] Suzuki H, Tamai N, Habu Y, Chang MO, Takaku H: Suppression of hepatitis C virus replication by baculovirus vector-mediated short-hairpin RNA expression. *FEBS Lett.,* 2008, 582:3085-3089.

[62] Suzuki H, Kaneko H, Tamai N, Miyano-Kurosaki N, Hashimoto K, Shimotohno K, et al: Suppression of HCV RNA replication by baculovirus mediated shRNA expression. *Nucleic Acids Symp. Ser.,* (Oxf) 2005, 339-340.

[63] Suzuki H, Saitoh H, Suzuki T, Takaku H: Inhibition of influenza virus by baculovirus-mediated shRNA. *Nucleic Acids Symp. Ser.,* (Oxf) 2009, 287-288.

INDEX

A

B

C

D

E

F

G

H

I

N

S

T

X

Y